Introduction to Digital Design
Using Digilent FPGA Boards
— Block Diagram / VHDL Examples

Richard E. Haskell
Darrin M. Hanna

Oakland University, Rochester, Michigan

LBE Books
Rochester Hills, MI

ISBN 978-0-9801337-6-9

Second Printing

Published by LBE Books, LLC
1202 Walton Boulevard
Suite 214
Rochester Hills, MI 48307

www.lbebooks.com

Preface

A major revolution in digital design has taken place over the past decade. Field programmable gate arrays (FPGAs) can now contain over a million equivalent logic gates and tens of thousands of flip-flops. This means that it is not possible to use traditional methods of logic design involving the drawing of logic diagrams when the digital circuit may contain thousands of gates. The reality is that today digital systems are designed by writing software in the form of hardware description languages (HDLs). The most common HDLs used today are VHDL and Verilog. Both are in widespread use. When using these hardware description languages the designer typically describes the *behavior* of the logic circuit rather than writing traditional Boolean logic equations. Computer-aided design tools are used to both *simulate* the VHDL or Verilog design and to *synthesize* the design to actual hardware.

This book assumes no previous knowledge of digital design. We use 30 examples to show you how to get started designing digital circuits that you can implement on a Xilinx Spartan3E FPGA using either the Digilent BASYS™ system board that can be purchased from www.digilentinc.com for $59 or the Digilent Nexys-2 board that costs $99. We will use Active-HDL from Aldec to design, simulate, synthesize, and implement our digital designs. A free student edition of Active-HDL is available from Aldec, Inc. (www.aldec.com). To synthesize your designs to a Spartan3E FPGA you will need to download the free ISE WebPACK from Xilinx, Inc. (www.xilinx.com). The Xilinx synthesis tools are called from within the Aldec Active-HDL integrated GUI. We will use the ExPort utility to download your synthesized design to the Spartan3E FPGA. ExPort is part of the Adept software suite that you can download free from Digilent, Inc. (www.digilentinc.com). A more complete book called *Digital Design Using Digilent FPGA Boards – VHDL / Active-HDL Edition* is also available from Digilent or LBE Books (www.lbebooks.com). This more comprehensive book contains over 75 examples including examples of using the VGA and PS/2 ports. Similar books that use Verilog are also available from Digilent or LBE Books.

Many colleagues and students have influenced the development of this book. Their stimulating discussions, probing questions, and critical comments are greatly appreciated.

Richard E. Haskell
Darrin M. Hanna

Introduction to Digital Design
Using Digilent FPGA Boards
— Block Diagram / VHDL Examples

Table of Contents

Introduction

Digital Design Using FPGAs

The first integrated circuits that were developed in the early 1960s contained less that 100 transistors on a chip and are called small-scale integrated (SSI) circuits. Medium-scale integrated (MSI) circuits, developed in the late 1960s, contain up to several hundreds of transistors on a chip. By the mid 1970s large-scale integrated (LSI) circuits containing several thousands of transistors had been developed. Very-large-scale integrated (VLSI) circuits containing over 100,000 transistors had been developed by the early 1980s. This trend has continued to the present day with 1,000,000 transistors on a chip by the late 1980s, 10,000,000 transistors on a chip by the mid-1990s, over 100,000,000 transistors by 2004, and up to 1,000,000,000 transistors on a chip today. This exponential growth in the amount of digital logic that can be packed into a single chip has produced serious problems for the digital designer. How can an engineer, or even a team of engineers, design a digital logic circuit that will end up containing millions of transistors?

In Appendix C we show that any digital logic circuit can be made from only three types of basic gates: AND, OR, and NOT. In fact, we will see that any digital logic circuit can be made using only NAND gates (or only NOR gates), where each NAND or NOR gate contains four transistors. These basic gates were provided in SSI chips using various technologies, the most popular being transistor-transistor logic (TTL). These TTL chips were the mainstay of digital design throughout the 1960s and 1970s. Many MSI TTL chips became available for performing all types of digital logic functions such as decoders, adders, multiplexers, comparators, and many others.

By the 1980s thousands of gates could fit on a single chip. Thus, several different varieties of *programmable logic devices* (PLDs) were developed in which arrays containing large numbers of AND, OR, and NOT gates were arranged in a single chip without any predetermined function. Rather, the designer could design any type of digital circuit and implement it by connecting the internal gates in a particular way. This is usually done by opening up fuse links within the chip using computer-aided tools. Eventually the equivalent of many PLDs on a single chip led to *complex programmable logic devices* (CPLDs).

Field Programmable Gate Arrays (FPGAs)

A completely different architecture was introduced in the mid-1980's that uses RAM-based lookup tables instead of AND-OR gates to implement combinational logic. These devices are called *field programmable gate arrays* (FPGAs). The device consists of an array of *configurable logic blocks* (CLBs) surrounded by an array of I/O blocks. The Spartan-3E from Xilinx also contains some blocks of RAM, 18 x 18 multipliers, as well as Digital Clock Manager (DCM) blocks. These DCMs are used to eliminate clock distribution delay and can also increase or decrease the frequency of the clock.

Each CLB in the Spartan-3E FPGA contains four slices, each of which contains two 16 x 1 RAM look-up tables (LUTs), which can implement any combinational logic function of four variables. In addition to two look-up tables, each slice contains two D flip-flops which act as storage devices for bits. The basic architecture of a Spartan-3E FPGA is shown in Fig. 1.

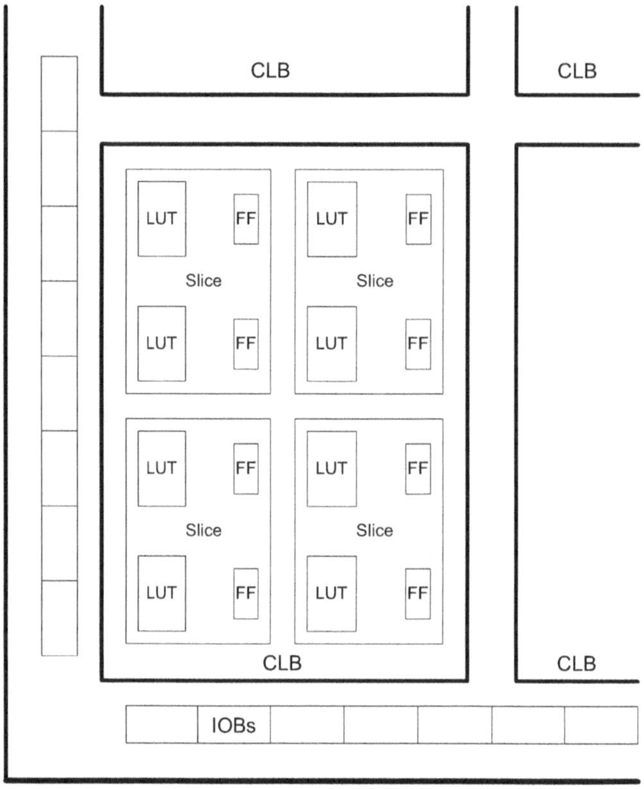

Figure 1 Architecture of a Spartan-3E FPGA

The BASYS board from Digilent contains a Xilinx Spartan3E-100 TQ144 FPGA. This chip contains 240 CLBs arranged as 22 rows and 16 columns. There are therefore 960 slices with a total of 1,920 LUTs and flip-flops. This part also contains 73,728 bits of block RAM. Half of the LUTs on the chip can be used for a maximum of 15,360 bits of distributed RAM.

By contrast the Nexys-2 board from Digilent contains a Xilinx Spartan3E-500 FG320 FPGA. This chip contains 1,164 CLBs arranged as 46 rows and 34 columns. There are therefore 4,656 slices with a total of 9,312 LUTs and flip-flops. This part also contains 368,640 bits of block RAM. Half of the LUTs on the chip can be used for a maximum of 74,752 bits of distributed RAM.

In general, FPGAs can implement much larger digital systems than CPLDs as illustrated in Table 1. The column labeled *No. of Gates* is really equivalent gates as we have seen that FPGAs really don't have AND and OR gates, but rather just RAM look-up tables. (Each slice does include two AND gates and two XOR gates as part of carry and arithmetic logic used when implementing arithmetic functions including addition and multiplication.) Note from Table 1 that FPGAs can have the equivalent of millions of gates and tens of thousands of flip-flops.

Table 1 Comparing Xilinx CPLDs and FPGAs

Xilinx Part	No. of Gates	No. of I/Os	No. of CLBs	No. of Flip-flops	Block RAM (bits)
CPLDs					
9500 family	800 – 6,400	34 – 192		36 - 288	
FPGAs					
Spartan	5,000 – 40,000	77 – 224	100 – 784	360 – 2,016	
Spartan II	15,000 – 200,000	86 – 284	96 – 1,176	642 – 5,556	16,384 – 57,344
Spartan IIE	23,000 – 600,000	182 – 514	384 – 3,456	2,082 – 15,366	32,768 – 294,912
Spartan 3	50,000 – 5,000,000	124 – 784	192 – 8,320	2,280 – 71,264	73,728 – 1,916,928
Spartan-3E	100,000 – 1,600,000	108 – 376	240 – 3,688	1,920 – 29,505	73,728 – 663,552
Virtex	57,906 – 1,124,022	180 – 512	384 – 6,144	2,076 – 26,112	32,768 – 131,072
Virtex E	71,693 – 4,074,387	176 – 804	384 – 16,224	1,888 – 66,504	65,536 – 851,968
Virtex-II	40,960 – 8,388,608	88 – 1,108	64 – 11,648	1,040 – 99,832	73,728 – 3,096,576

Modern Design of Digital Systems

The traditional way of designing digital circuits is to draw logic diagrams containing SSI gates and MSI logic functions. However, by the late 1980s and early 1990s such a process was becoming problematic. How can you draw schematic diagrams containing hundreds of thousands or millions of gates? As programmable logic devices replaced TTL chips in new designs a new approach to digital design became necessary. Computer-aided tools are essential to designing digital circuits today. What has become clear over the last decade is that today's digital engineer designs digital systems by writing software! This is a major paradigm shift from the traditional method of designing digital systems. Many of the traditional design methods that were important when using TTL chips are less important when designing for programmable logic devices.

Today digital designers use *hardware description languages* (HDLs) to design digital systems. The most widely used HDLs are VHDL and Verilog. Both of these hardware description languages allow the user to design digital systems by writing a program that describes the behavior of the digital circuit. The program can then be used to both *simulate* the operation of the circuit and *synthesize* an actual implementation of the circuit in a CPLD, an FPGA, or an application specific integrated circuit (ASIC).

Another recent trend is to design digital circuits using block diagrams or graphic symbols that represent higher-level design constructs. These block diagrams can then be *compiled* to produce Verilog or VHDL code. We will illustrate this method in this book.

We will use Active-HDL from Aldec for designing our digital circuits. This integrated tool allows you to enter your design using either a block diagram editor (BDE) or by writing Verilog or VHDL code using the hardware description editor (HDE). Once your hardware has been described you can use the functional simulator to produce waveforms that will verify your design. This hardware description can then be synthesized to logic equations and implemented or mapped to the FPGA architecture.

We include a tutorial for using Active-HDL in Appendix A. A free student version of Active-HDL is available on their website.[1] We will use Xilinx ISE for synthesizing our VHDL designs. You can download a free version of ISE™ WebPACK™ from the Xilinx website.[2] This WebPACK™ synthesis tool can be run from within the Aldec Active-HDL development environment as shown in the tutorial in Appendix A. The implementation process creates a *.bit* file that is downloaded to a Xilinx FPGA on the BASYS board or Nexys-2 shown in Fig. 2. The BASYS board is available to students for $59 from Digilent, Inc.[3] This board includes a 100k-gate equivalent Xilinx Spartan3E FPGA (250k-gate capacity is also available), 8 slide switches, 4 pushbutton switches, 8 LEDs, and four 7-segment displays. The frequency of an on-board clock can be set to 25 MHz, 50 MHz, or 100 MHz using a jumper. There are connectors that allow the board to be interfaced to external circuits. The board also includes a VGA port and a PS2 port. The use of these ports are described in a different book.[4] Another more advanced board, the Nexys-2 board, is also available to students for $99 from Digilent. The Nexys-2 board is similar to the BASYS board except that it contains a 500k- or 1200k-gate equivalent Spartan 3E FPGA, a Hirose FX2 interface for additional add-on component boards, 16 MB of cellular RAM, 16 MB of flash memory, a 50 MHz clock and a socket for a second oscillator. The Nexys-2 is ideally suited for embedded processors.

All of the examples in this book can be used on both the BASYS board and the Nexys-2 board. The only difference is that you would use the file *basys2.ucf* to define the pinouts on the BASYS board and you would use the file *nexys2.ucf* to define the pinouts on the Nexys-2 board. Both of these files are available to download from www.lbebooks.com. Table 2 shows the jumper settings you would use on the two boards.

(a) (b)

Figure 2 (a) BASYS board, (b) Nexys-2 Board

---

[1] http://www.aldec.com/education/

[2] http://www.xilinx.com

[3] http://www.digilentinc.com

[4] *Digital Design Using Digilent FPGA Boards – VHDL / Active-HDL Edition*; available from www.lbebooks.com.

Table 1.2 Board Jumper Settings

BASYS Board	Nexys-2 Board
Set the JP3 jumper to JTAG	Set the POWER SELECT jumper to USB
Remove the JP4 jumper to select a 50 MHz clock	Set the MODE jumper to JTAG

VHDL

VHDL is based on the Ada software programming language but it is *not* Ada nor is it a software programming language. VHDL is a *hardware description language* that is designed to model digital logic circuits. It simply has syntax similar to the Ada programming language but the way it behaves is different. In this book you will learn VHDL by studying the examples we use to describe digital logic and then doing some of the VHDL problems at the end of each chapter.

In this book we begin by using the Active-HDL block diagram editor to draw logic circuits using basic gates. When you *compile* these block diagrams Active-HDL will generate the corresponding VHDL code. The block diagram representing your logic circuit can then be used as a module in a higher-level digital design. This higher-level design can then be compiled to produce its corresponding VHDL code. This hierachical block diagram editor will make it easy to design top-level designs.

Sometimes it will be easier to design a digital module by writing a VHDL program directly rather than drawing it using gates. When you do this you can still use the block diagram for this module in higher-level designs. We will illustrate this process in many of our examples.

Just like any programming language, you can only learn VHDL by actually writing VHDL programs and simulating the designs using a VHDL simulator that will display the waveforms of the signals in your design. This is a good way to learn not only VHDL but digital logic as well.

A companion book[5] that uses Verilog instead of VHDL is available from www.digilentinc.com or www.lbebooks.com. More comprehensive Verilog and VHDL books are also available.[6,7]

---

[5] *Introduction to Digital Design Usign Digilent FPGA Boards – Block Diagram/Verilog Examples*

[6] *Digital Design Using Digilent FPGA Boards – Verilog / Active-HDL Edition*, LBE Books, 2009.

[7] *Digital Design Using Digilent FPGA Boards – VHDL / Active-HDL Edition*, LBE Books, 2009.

Example 1

Switches and LEDs

In this example we will show the basic structure of a VHDL program and how to write logic equations for 2-input gates. Example 1a will show the simulation results using Aldec Active-HDL and Example 1b will show how to synthesize the program to a Xilinx FPGA on the BASYS or Nexys-2 board.

Prerequisite knowledge:
None
Learned in this Example:
Use of Aldec Active-HDL – Appendix A

1.1 Slide Switches

The slide switches on the BASYS and Nexys-2 boards are connected to pins on the FPGA through a resistor *R* as shown in Fig. 1.1. The value of *R* is 4.7 kΩ on the BASYS board and 10 kΩ on the Nexys-2 board. When the slide switch is down it is connected to ground and the input *sw*(*i*) to the FPGA is read as a logic 0. When the slide switch is up it is connected to 3.3 V and the input *sw*(*i*) to the FPGA is read as a logic 1.

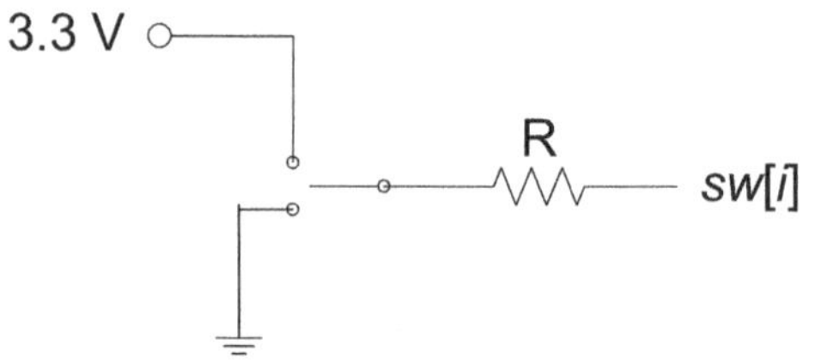

Figure 1.1 Slide switch connection

There are eight slide switches on the BASYS and Nexys-2 boards. The eight pin numbers on the FPGA corresponding to the eight slide switches are given in a *.ucf* file. The file *basys2.ucf* shown in Listing 1.1 defines the pin numbers for all I/O on the BASYS board. Note that we have named the slide switches *sw*(*i*), *i* = 0:7, which correspond to the switch labels on the board. We will always name the slide switches *sw*(*i*) in our top-level designs so that we can use the *basys2.ucf* file without change. Because the pin numbers on the Nexys-2 board are different from those on the BASYS board we will use a different file called *nexys2.ucf* to define the pin numbers on the Nexys-2 board. The names of the I/O ports, however, will be the same for both boards. Therefore, all of the examples in this book can be used with either board by simply using the proper *.ucf* file when implementing the design. Both of these *.ucf* files can be downloaded from www.lbebooks.com.

1.2 LEDs

A light emitting diode (LED) emits light when current flows through it in the positive direction as shown in Fig. 1.2. Current flows through the LED when the voltage on the *anode* side (the wide side of the black triangle) is made higher than the voltage on the *cathode* side (the straight line connected to the apex of the black triangle). When current flows through a lighted LED the forward voltage across the LED is typically between +1.5 and +2.0 volts. If voltage *V2* in Fig. 1.2 is less than or equal to voltage *V1* then no current can flow through the LED and therefore no light will be emitted. If voltage *V2* is greater than voltage *V1* then current will flow through the resistor *R* and the LED. The resistor is used to limit the amount of current that flows through the LED. Typical currents needed to light LEDs range from 2 to 15 milliamps.

Listing 1.1 basys2.ucf

```
# Pin assignment for LEDs
NET "ld<7>" LOC = "p2" ;
NET "ld<6>" LOC = "p3" ;
NET "ld<5>" LOC = "p4" ;
NET "ld<4>" LOC = "p5" ;
NET "ld<3>" LOC = "p7" ;
NET "ld<2>" LOC = "p8" ;
NET "ld<1>" LOC = "p14" ;
NET "ld<0>" LOC = "p15" ;

# Pin assignment for slide switches
NET "sw<7>" LOC = "p6";
NET "sw<6>" LOC = "p10";
NET "sw<5>" LOC = "p12";
NET "sw<4>" LOC = "p18";
NET "sw<3>" LOC = "p24";
NET "sw<2>" LOC = "p29";
NET "sw<1>" LOC = "p36";
NET "sw<0>" LOC = "p38";

# Pin assignment for pushbutton switches
NET "btn<3>" LOC = "p41";
NET "btn<2>" LOC = "p47";
NET "btn<1>" LOC = "p48";
NET "btn<0>" LOC = "p69";

# Pin assignment for 7-segment displays
NET "a_to_g<6>"  LOC = "p25"  ;
NET "a_to_g<5>"  LOC = "p16"  ;
NET "a_to_g<4>"  LOC = "p23"  ;
NET "a_to_g<3>"  LOC = "P21"  ;
NET "a_to_g<2>"  LOC = "p20"  ;
NET "a_to_g<1>"  LOC = "p17"  ;
NET "a_to_g<0>"  LOC = "p83"  ;
NET "dp"  LOC = "p22"  ;

NET "an<3>" LOC = "p26";
NET "an<2>" LOC = "p32";
NET "an<1>" LOC = "p33";
NET "an<0>" LOC = "p34";

# Pin assignment for clock
NET "mclk" LOC = "p54";
```

There are two different ways that an I/O pin of an FPGA can be used to turn on an LED. The first is to connect the FPGA pin to *V2* in Fig. 1.2 and to connect *V1* to ground. Bringing the pin (*V2*) high will then turn on the LED. To turn off the LED the output pin would be brought low. This is the method used for the LEDs *ld*(7) – *ld*(0) on the BASYS and Nexys-2 boards.

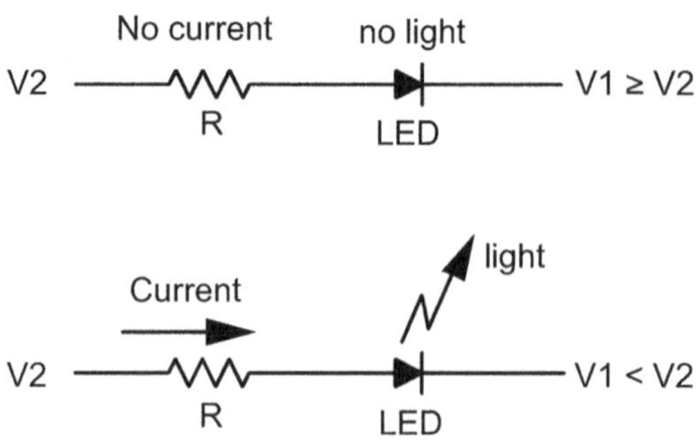

Figure 1.2 Turning on an LED

The second method is to connect the FPGA pin to *V1* in Fig. 1.2 and to connect *V2* to a constant voltage. Bringing the pin (*V1*) low will then turn on the LED. To turn off the LED the output pin would be brought high. This voltage should be equal to *V2* to make sure no current flows through the LED. This second method is the method used for the 7-segment displays on the BASYS and Nexys-2 boards. Examples 9 and 10 will show how to display hex digits on the 7-segment displays.

1.3 Connecting the Switches to the LEDs

Part 1 of the tutorial in Appendix A shows how to connect the input switches to the output LEDs using the block diagram editor (BDE) in Active-HDL. The result is shown in Fig. 1.3.

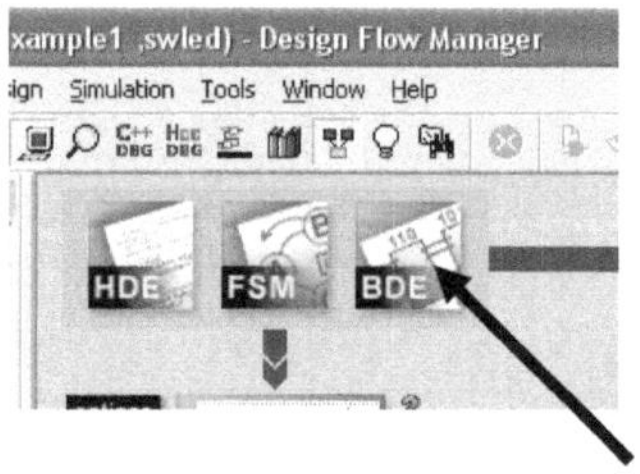

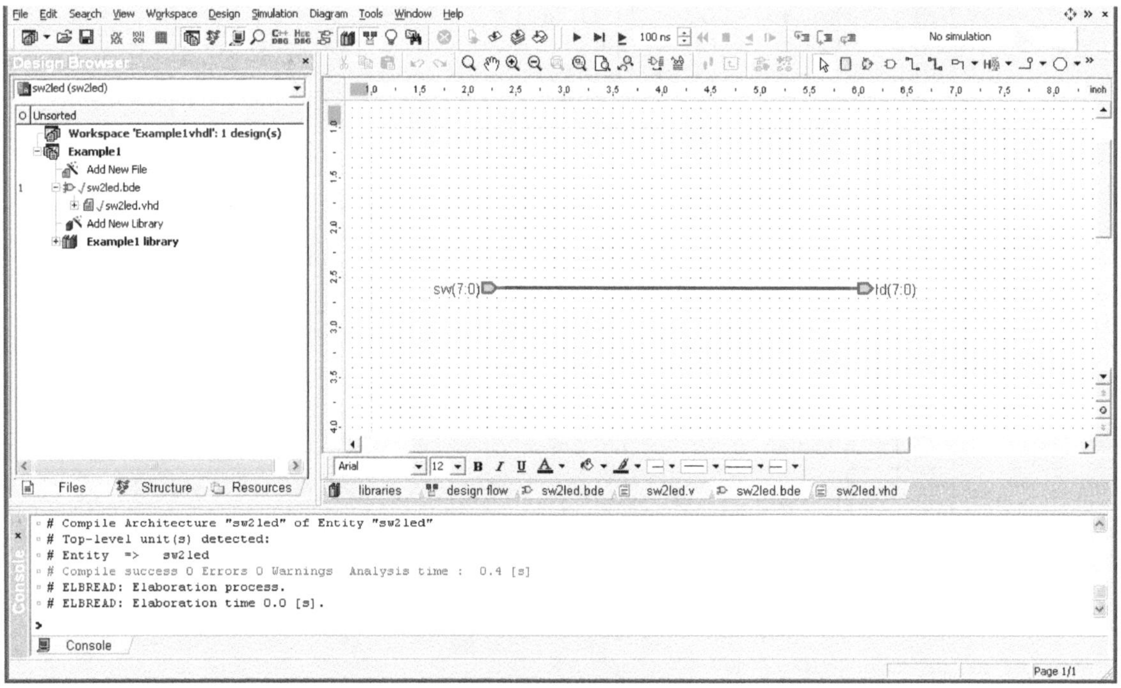

Figure 1.3 Connecting the eight switches to the eight LEDs

Compiling the file *sw2led.bde* generates the VHDL file *sw2led.vhd* shown in Listing 1.2. Alternatively, by selecting the hardware description editor (HDE) the entity and architecture declarations are automatically generated but you will need to write your own assignment statements. This can lead to the simpler VHDL program shown in Listing 1.3 where we can write a single assignment statement using the assignment operator, <=, to replace the two intermediate assignment statements in Listing 1.2. It is unnecessary to define the intermediate bus *BUS23*(7:0).

Listing 1.2 sw2led.vhd

```
library IEEE;
use IEEE.std_logic_1164.all;


entity sw2led is
  port(
        sw : in STD_LOGIC_VECTOR(7 downto 0);
        ld : out STD_LOGIC_VECTOR(7 downto 0)
  );
end sw2led;

architecture sw2led of sw2led is

---- Signal declarations used on the diagram ----

signal BUS23 : STD_LOGIC_VECTOR (7 downto 0);

begin

---- Terminal assignment ----

    -- Inputs terminals
      BUS23 <= sw;

    -- Output\buffer terminals
      ld <= BUS23

end sw2led;
```

Listing 1.3 sw2led2.vhd

```
library IEEE;
use IEEE.std_logic_1164.all;

entity sw2led2 is
  port(
        sw : in STD_LOGIC_VECTOR(7 downto 0);
        ld : out STD_LOGIC_VECTOR(7 downto 0)
  );
end sw2led2;

architecture sw2led2 of sw2led2 is
begin
      ld <= sw;
end sw2led2;
```

Note in the entity in Listing 1.3 that the input *sw* and the output *ld* are defined to be of type STD_LOGIC_VECTOR (7 *downto* 0). For simulation purposes this type is defined to have nine possible values. In addition to the usual 0 and 1 the other seven possible values are U (uninitialized), X (unknown), Z (high impedance), W (weak unknown), L (weak 0), H (weak 1), and – (don't care).

In Parts 2 and 3 of the tutorial in Appendix A we show how to synthesize, implement, and download the design to the FPGA board. In summary, the steps you follow to implement a digital design on the BASYS or Nexys-2 board are the following:

1. Create a new project and design name.
2. Using the BDE create a logic diagram.
3. Save and compile the *.bde* file.
4. Optionally simulate the design (see Example 2).
5. Synthesize the design selecting the Spartan3E family and the 3s100etq144 device for the BASYS board and the 3s500efg320 device for the Nexys-2 board.
6. Implement the design using either *basys2.ucf* or *nexys2.ucf* as the custom constraint file. Check *Allow Unmatched LOC Constraints* under *Translate* and uncheck *Do Not Run Bitgen* under *BitStream*. Select *JTAG Clock* as the start-up clock under *Startup Options*.
7. Use *ExPort* to download the *.bit* file to the FPGA board.

At this point the switches are connected to the LEDs. Turning on a switch will light up the corresponding LED.

Problem

1.1 The four pushbuttons on the BASYS and Nexys-2 boards are connected to pins on the FPGA using the circuit shown in Fig. 1.4. The value of *R* is 4.7 kΩ on the BASYS board and 10 kΩ on the Nexys-2 board. When the pushbutton is up the two resistors pull the input down to ground and the input *btn*(*i*) to the FPGA is read as a logic 0. When the pushbutton is pressed the input is pulled up to 3.3 V and the input *btn*(*i*) to the FPGA is read as a logic 1. Create a *.bde* file using Active-HDL that will connect the four pushbuttons to the rightmost four LEDs. Compile and implement the program. Download the *.bit* file to the FPGA board and test it by pressing the pushbuttons.

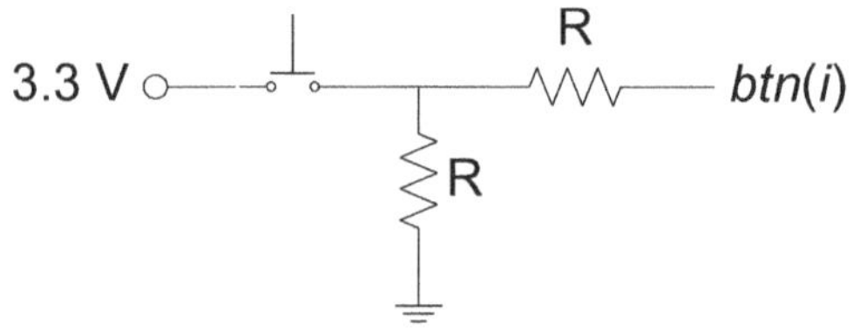

Figure 1.4 Pushbutton connection

Example 2

2-Input Gates

In this example we will design a circuit containing six different 2-input gates. Example 2a will show the simulation results using Aldec Active-HDL and Example 2b will show how to synthesize the program to a Xilinx FPGA on a Digilent board.

Prerequisite knowledge:

Appendix C – Basic Logic Gates
Appendix A – Use of Aldec Active-HDL

2.1 Generating the Design File *gates2.bde*

Part 4 of the tutorial in Appendix A shows how to connect two inputs *a* and *b* to the inputs of six different gates using the block diagram editor (BDE) in Active-HDL. The result is shown in Fig. 2.1. Note that we have named the outputs of the gates the name of the gate including an underscore. Identifier names in VHDL can contain any letter, digit, underscore _, or $. The identifier can not begin with a digit or be a VHDL keyword. VHDL is *not* case sensitive.

The name of this file is *gates2.bde*. When you compile this file the VHDL program *gates2.vhd* shown in Listing 2.1 is generated.

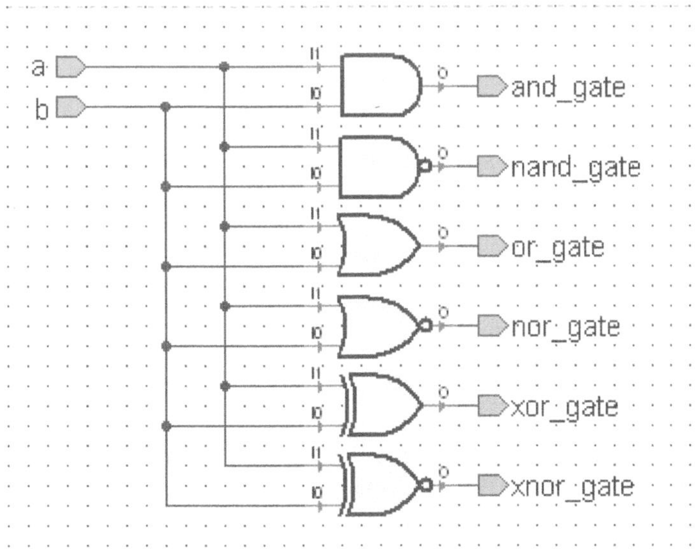

Figure 2.1 Circuit diagram for Example 2

Listing 2.1 gates2.vhd

```
-- Example 2a: gates2
library IEEE;
use IEEE.std_logic_1164.all;

entity gates2 is
  port(
        a : in STD_LOGIC;
        b : in STD_LOGIC;
        and_gate : out STD_LOGIC;
        nand_gate : out STD_LOGIC;
        nor_gate : out STD_LOGIC;
        or_gate : out STD_LOGIC;
        xnor_gate : out STD_LOGIC;
        xor_gate : out STD_LOGIC
  );
end gates2;

architecture gates2 of gates2 is
begin
----  Component instantiations  ----
      and_gate <= b and a;
      nand_gate <= not(b and a);
      or_gate <= b or a;
      nor_gate <= not(b or a);
      xor_gate <= b xor a;
      xnor_gate <= not(b xor a);

end gates2;
```

The logic diagram in Fig. 2.1 contains six different gates. This logic circuit is described by the VHDL program shown in Listing 2.1. The first line in Listing 2.1 is a comment. Comments in VHDL follow the double dash --. All VHDL programs begin with an *entity* statement containing the name of the entity (*gates2* in this case) followed by a list of all input and output signals together with their direction and type. We will generally use lower case names for signals. The direction of the input and output signals is given by the VHDL statements *in*, *out*, or *inout* (for a bi-directional signal).

To describe the output of each gate in Fig. 2.1 we simply write the logic equation for that gate preceded by the assignment operator, <=. These are *concurrent* assignment statements which means that the statements can be written in any order.

2.2 Simulating the Design *gates2.bde*

Part 4 of the tutorial in Appendix A shows how to simulate this VHDL program using Active-HDL. The simulation produced in Appendix A is shown in Fig. 2.2. Note that the waveforms shown in Fig. 2.2 verify the truth tables for the six gates. Also note that two clock stimulators were used for the inputs *a* and *b*. By making the period of the clock stimulator for the input *a* twice the period of the clock stimulator for the input *b* all four combinations of the inputs *a* and *b* will be generated in one period of the input *a*.

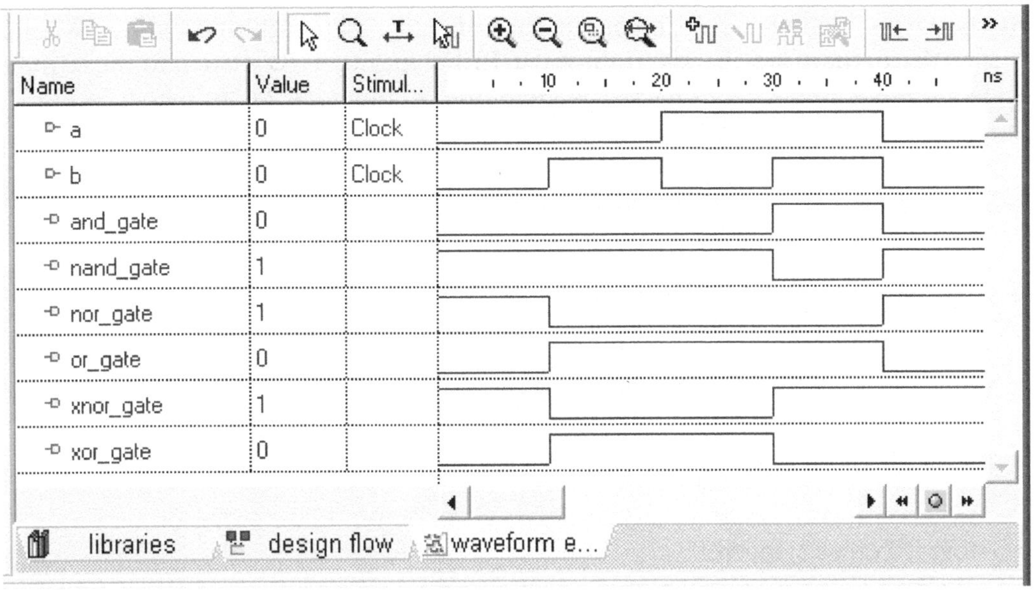

Figure 2.2 Simulation of logic circuit in Fig. 2.1

2.3 Generating a Top-Level Design

Part 5 of the tutorial in Appendix A shows how to create a top-level design for the *gates2* circuit. In order to use the constraint files *basys2.ucf* or *nexys2.ucf* described in Example 1 we must name the switch inputs *sw*(*i*) and the LED outputs *ld*(*i*). This top-level design, as created in Part 5 of Appendix A is shown in Fig. 2.3. The module *gates2* in Fig. 2.3 contains the logic circuit shown in Fig. 2.1. Note that each wire connected to a bus must be labeled to identify its connection to the bus lines.

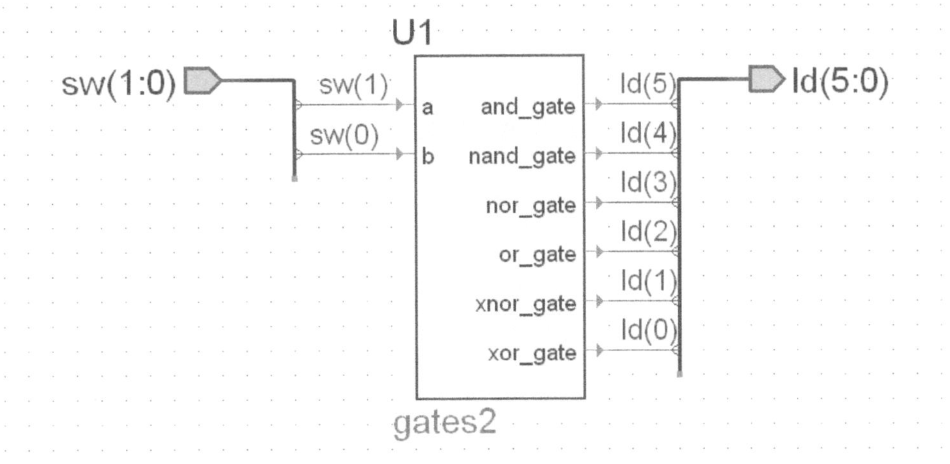

Figure 2.3 Top-level design for Example 2

Compiling the top-level design shown in Fig. 2.3 will generate the VHDL program shown in Listing 2.2. The inputs are now the two rightmost slide switches, *sw*(1:0), and the outputs are the six right-most LEDs *ld*(5:0). To associate these inputs and outputs with the inputs *a* and *b* and the six output in the *gates2* component in Fig. 2.1 and Listing 2.1 we use the VHDL *port map statement*

```
U1 : gates2
  port map(
       a => sw(1),
       b => sw(0),
       and_gate => ld(5),
       nand_gate => ld(4),
       nor_gate => ld(3),
       or_gate => ld(2),
       xnor_gate => ld(1),
       xor_gate => ld(0)
  );
```

Listing 2.2 gates2_top.vhd

```
-- Example 2b: gates2_top
library IEEE;
use IEEE.std_logic_1164.all;
library EXAMPLE2;

entity gates2_top is
  port(
       sw : in STD_LOGIC_VECTOR(1 downto 0);
       ld : out STD_LOGIC_VECTOR(5 downto 0)
  );
end gates2_top;

architecture gates2_top of gates2_top is
component gates2
  port (
       a : in std_logic;
       and_gate : out std_logic;
       b : in std_logic;
       nand_gate : out std_logic;
       nor_gate : out std_logic;
       or_gate : out std_logic;
       xnor_gate : out std_logic;
       xor_gate : out std_logic
  );
end component;

begin
U1 : gates2
  port map(
       a => sw(1),
       b => sw(0),
       and_gate => ld(5),
       nand_gate => ld(4),
       nor_gate => ld(3),
       or_gate => ld(2),
       xnor_gate => ld(1),
       xor_gate => ld(0)
  );
end gates2_top;
```

This VHDL *port map* statement begins with an arbitrary name for the component in the top-level design. Here we call it U1. This is followed by the name of the component being instantiated, in this case *gates2* from Listing 2.1. Then using the *port map* statement enclosed in parentheses are the inputs and outputs from Listing 2.1 associated with corresponding inputs and outputs in the top-level design in Fig. 2.3. Note that we connect the input *a* in Listing 2.1 to the input *sw*(1) on the FPGA board. The input *b* in Listing 2.1 is connected to *sw*(0) and the outputs *and_gate*, *nand*_gate, *or_gate*, *nor_gate*, *xor_gate*, and *xnor_gate* are connected to the corresponding LED outputs *ld*(5:0). These associations can be made in this way in any order. The *port map* statement in Listing 2.2 generated from the top-level block diagram are associated in alphabetical order.

Follow the steps in the tutorial in Appendix A and implement this design on the FPGA board. Note that when you change the settings of the two right-most slide switches the LEDs will indicate the outputs of the six gates.

Example 3

Multiple-Input Gates

In this example we will design a circuit containing multiple-input gates. We will create a logic circuit containing 4-input AND, OR, and XOR gates. We will leave it as a problem for you to create a logic circuit containing 4-input NAND, NOR, and XNOR gates.

Prerequisite knowledge:
Appendix C – Basic Logic Gates
Appendix A – Use of Aldec Active-HDL

3.1 Behavior of Multiple-Input Gates

The AND, OR, NAND, NOR, XOR, and XNOR gates we studied in Example 1 had two inputs. The basic definitions hold for multiple inputs. A multiple-input AND gate is shown in Fig. 3.1. *The output of an AND gate is HIGH only if all inputs are HIGH*. To describe this multiple-input AND gate in VHDL we could simply write the logic equation as

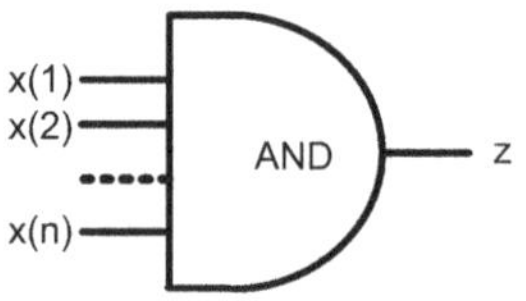

Figure 3.1
Multiple-input AND gate.

z <= x(1) **and** x(2) **and** ... **and** x(n);

A multiple-input OR gate is shown in Fig. 3.2. *The output of an OR gate is LOW only if all inputs are LOW*. Just As with the AND gate we can write the logic equation as

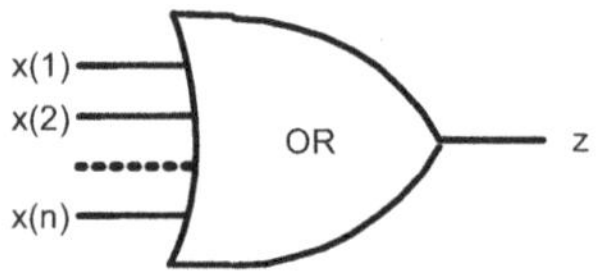

Figure 3.2
Multiple-input OR gate.

```
z <= x(1) or x(2) or ... or x(n);
```

A multiple-input NAND gate is shown in Fig. 3.3. *The output of a NAND gate is LOW only if all inputs are HIGH*. We can write the logic equation as

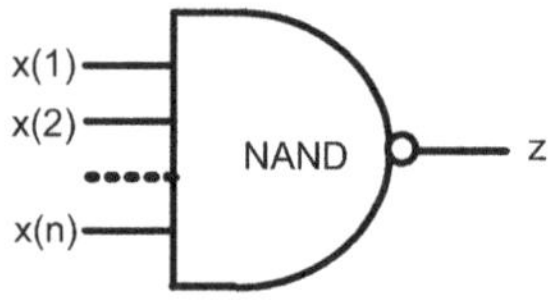

Figure 3.3
Multiple-input NAND gate.

```
z <= not(x(1) and x(2) and ... and x(n));
```

A multiple-input NOR gate is shown in Fig. 3.4. *The output of a NOR gate is HIGH only if all inputs are LOW*. We can write the logic equation as

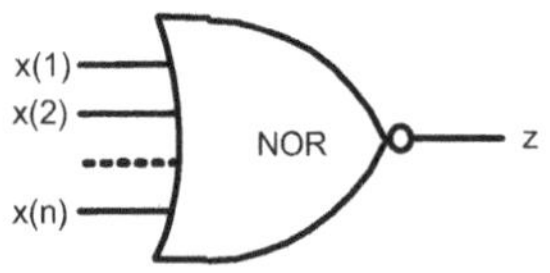

Figure 3.4
Multiple-input NOR

```
z <= not(x(1) or x(2) or ... or x(n));
```

A multiple-input XOR gate is shown in Fig. 3.5. What is the meaning of this multiple-input gate? Following the methods we used for the previous multiple-input gates we can write the logic equation as

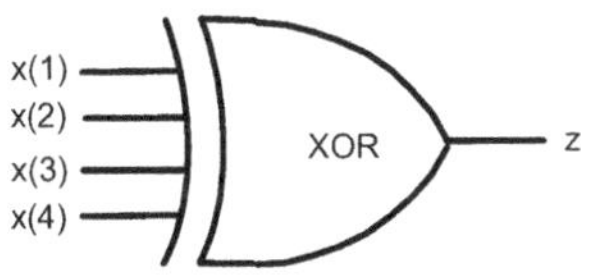

Figure 3.5
Multiple-input XOR gate.

.

```
z <= x(1) xor x(2) xor ... xor x(n);
```

We will create a 4-input XOR gate in this example to determine its meaning but first consider the multiple-input XNOR gate shown in Fig. 3.6. What is the meaning of this multiple-input gate? (See Problem 3.1 at the end of this example for the answer.) Following the methods we used for the previous multiple-input gates we can write the logic equation as

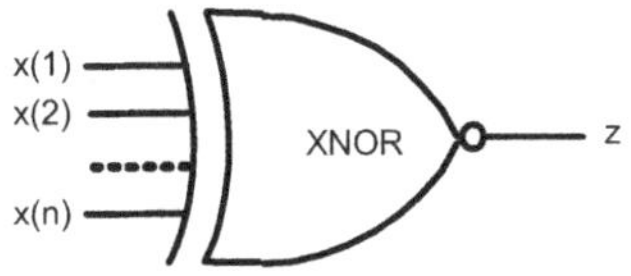

Figure 3.6
Multiple-input XNOR gate.

.

```
z <= not(x(1) xor x(2) xor ... xor x(n));
```

or we can use the following gate *instantiation statement* for an XNOR gate.

```
z <= x(1) xnor x(2) xnor ... xnor x(n);
```

3.2 Generating the Design File *gates4.bde*

Use the block diagram editor (BDE) in Active-HDL to create the logic circuit called *gates4.bde* shown in Fig. 3.7. A simulation of this circuit is shown in Fig. 3.8. From this simulation we see that <u>*the output of an XOR gate is HIGH only if the number of HIGH inputs is ODD*</u>.

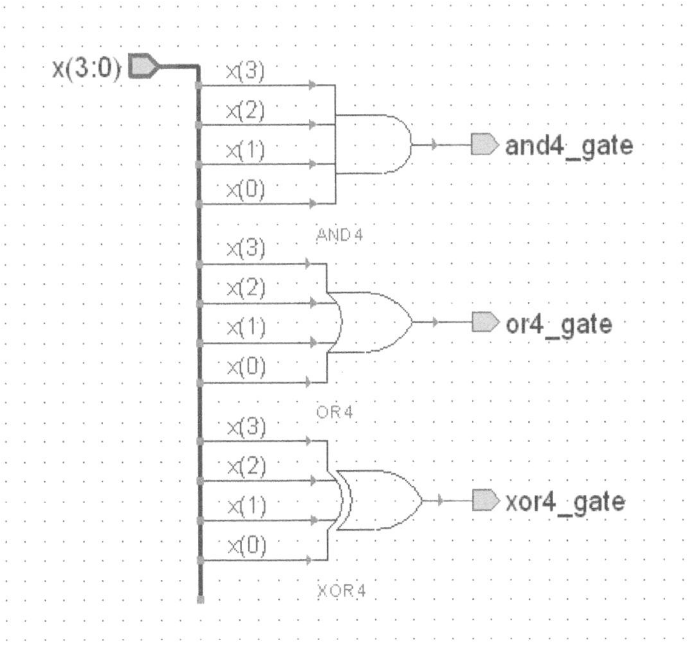

Figure 3.7 Block diagram for *gates4.bde*

If you look at the file *gates4.vhd* that is generated when you compile *gates4.bde* you will see that Active-HDL defines separate components for the 4-input AND, OR, and XOR gates and then uses a VHDL instantiation and *port map* statement to "wire" them together.

Alternatively, we could use the HDE editor to write the simpler VHDL program called *gates4b.vhd* shown in Listing 3.1 that uses standard VHDL logical operators to implement the three 4-input gates. This VHDL program will produce the same simulation as shown in Fig. 3.8.

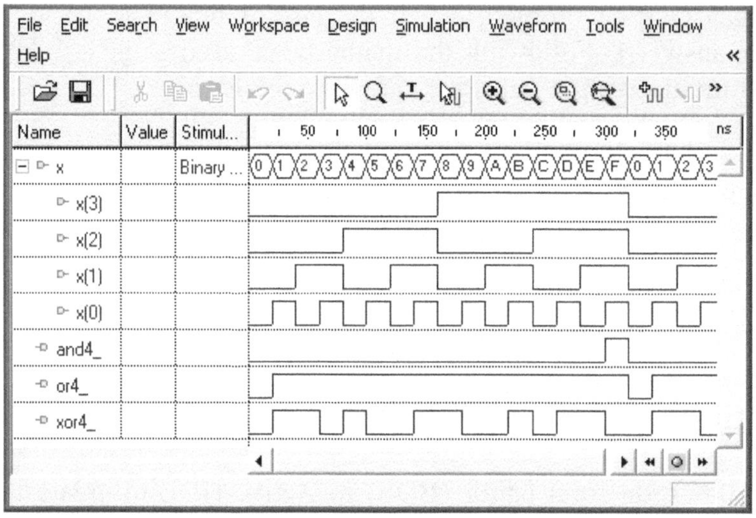

Figure 3.8 Simulation of the design *gates4.bde* shown in Fig. 3.7

Listing 3.1: gates4b.vhd

```
--Example 3: 4-input gates
library IEEE;
use IEEE.STD_LOGIC_1164.all;

entity gates4b is
       port(
             x : in STD_LOGIC_VECTOR(4 downto 1);
             and4_gate : out STD_LOGIC;
             or4_gate : out STD_LOGIC;
             xor4_gate : out STD_LOGIC
           );
end gates4b;

architecture gates4b of gates4b is
begin
  and4_gate <= x(1) and x(2) and x(3) and x(4);
  or4_gate <= x(1) or x(2) or x(3) or x(4);
  xor4_gate <= x(1) xnor x(2) xnor x(3) xnor x(4);
end gates4;
```

3.3 Generating the Top-Level Design *gates4_top.bde*

Fig. 3.9 shows the block diagram of the top-level design *gates4_top.bde*. The module *gates4* shown in Fig. 3.9 contains the logic circuit shown in Fig. 3.4. If you compile *gates4_top.bde* the VHDL program *gates4_top* shown in Listing 3.2 will be generated. Compile, synthesize, implement, and download this design to the FPGA board.

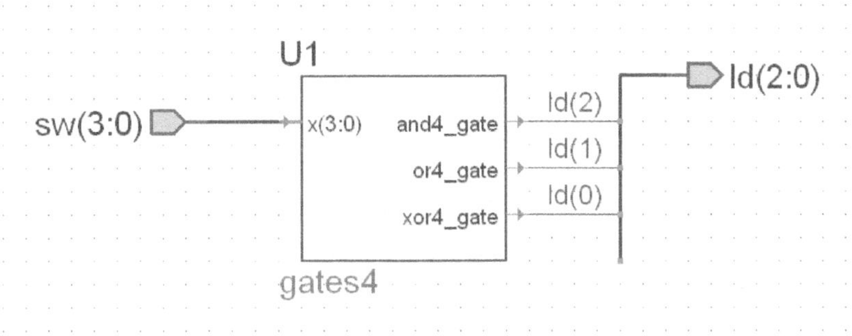

Figure 3.9 Block diagram for the top-level design *gates4_top.bde*

Listing 3.2: gates4_top.v

```
-- Example 2: 4-input gates - top level
library IEEE;
use IEEE.std_logic_1164.all;
library EXAMPLE3;

entity gates4_top is
  port(
       sw : in std_logic_vector(3 downto 0);
       ld : out STD_LOGIC_VECTOR(2 downto 0)
  );
end gates4_top;

architecture gates4_top of gates4_top is

component gates4
  port (
       x : in std_logic_vector(3 downto 0);
       and4_gate : out std_logic;
       or4_gate : out std_logic;
       xor4_gate : out std_logic
  );
end component;

begin
U1 : gates4
  port map(
       and4_gate => ld(2),
       or4_gate => ld(1),
       x => sw,
       xor4_gate => ld(0)
  );
end gates4_top;
```

Problem

3.1 Use the BDE to create a logic circuit containing 4-input NAND, NOR, and XNOR gates. Simulate your design and verify that *<u>the output of an XNOR gate is HIGH only if the number of HIGH inputs is EVEN</u>*. Create a top-level design that connects the four inputs to the rightmost four slide switches and the three outputs to the three rightmost LEDs. Implement your design and download it to the FPGA board.

3.2 The circuit shown at the right is for a 2 x 4 decoder. Use the BDE to create this circuit and simulate it using Active-HDL. Choose a counter stimulator for *x*(1:0) that counts every 20 ns, set *en* to a forced value of 1, and simulate it for 100 ns. Make a truth table with (*x*(1), *x*(0)) as the inputs and *y*(0:3) as the outputs. What is the behavior of this decoder?

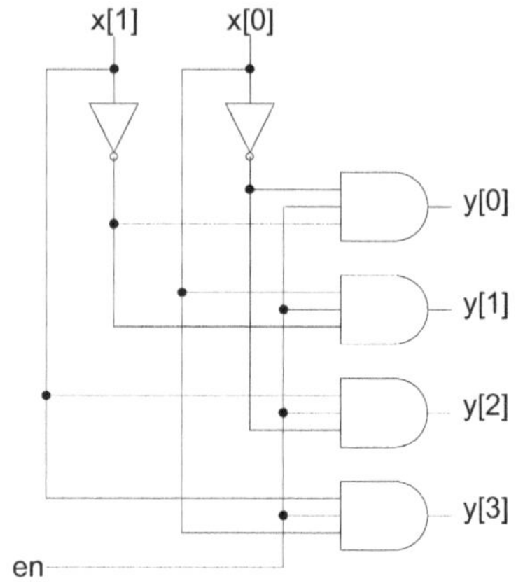

Example 4

Equality Detector

In this example we will design a 2-bit equality detector using two NAND gates and an AND gate.

Prerequisite knowledge:
Appendix C – Basic Logic Gates
Appendix A – Use of Aldec Active-HDL

4.1 Generating the Design File *eqdet2.bde*

The truth table for a 2-input XNOR gate is shown in Fig. 4.1. Note that the output *z* is 1 when the inputs *x* and *y* are equal. Thus, the XNOR gate can be used as a 1-bit equality detector.

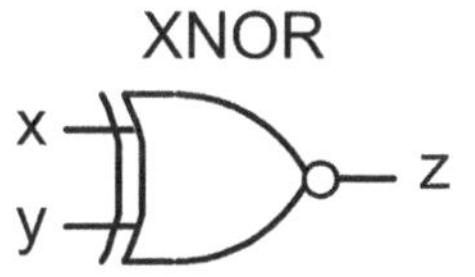

z = ~(x ^ y)

x	y	z
0	0	1
0	1	0
1	0	0
1	1	1

Figure 4.1 The XNOR gate is a 1-bit equality detector

By using two XNOR gates and an AND gate we can design a 2-bit equality detector as shown in Fig. 4.2. Use the BDE to create the file *eqdet2.bde* using Active-HDL.

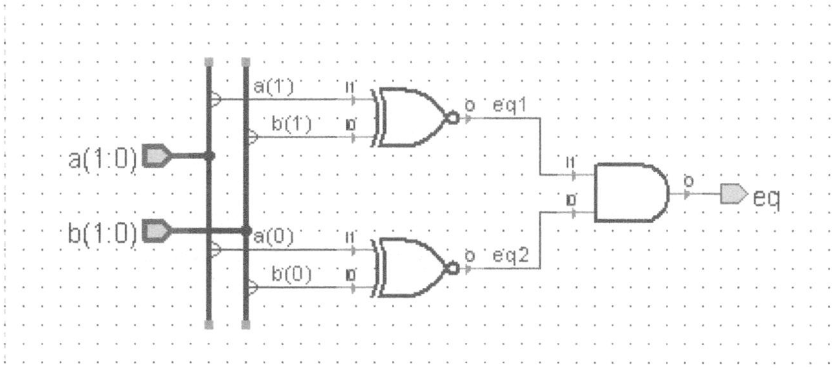

Figure 4.2 Block diagram of a 2-bit equality detector, *eqdet2.bde*

If you compile the file *eqdet2.bde* Active-HDL will generate the VHDL program *eqdet2.vhd* shown in Listing 4.1. A simulation of *eqdet2.bde* is shown in Fig. 4.3. Note that the output *eq* is 1 only if *a*(1:0) is equal to *b*(1:0).

Listing 4.1: eqdet2.vhd

```
-- Title          : eqdet2
library IEEE;
use IEEE.std_logic_1164.all;

entity eqdet2 is
  port(
        a : in STD_LOGIC_VECTOR(1 downto 0);
        b : in STD_LOGIC_VECTOR(1 downto 0);
        eq : out STD_LOGIC
  );
end eqdet2;

architecture eqdet2 of eqdet2 is
signal eq1 : STD_LOGIC;
signal eq2 : STD_LOGIC;

begin

eq1 <= not(b(1) xor a(1));
eq2 <= not(b(0) xor a(0));
eq <= eq2 and eq1;

end eqdet2;
```

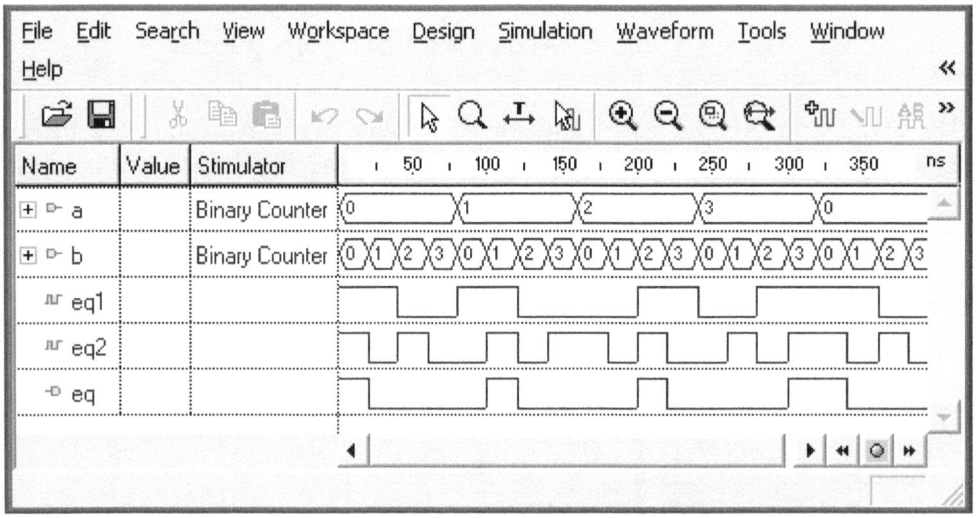

Figure 4.3 Simulation of the 2-bit equality detector, *eqdet2.bde*

Create a top-level design called *eqdet2_top.bde* that connects *a*(1:0) and *b*(1:0) to the rightmost four slide switches and connects the output *eq* to *ld*(0). Implement your design and download it to the FPGA board.

Example 5

2-to-1 Multiplexer: *if* Statement

In this example we will show how to design a 2-to-1 multiplexer and will introduce the VHDL *if* statement. Section 5.1 will define a multiplexer and derive the logic equations for a 2-to-1 multiplexer. Section 5.2 will illustrate the use of two versions of the VHDL *if* statement.

Prerequisite knowledge:

Karnaugh Maps – Appendix D
Use of Aldec Active-HDL – Appendix A

5.1 Multiplexers

An *n*-input multiplexer (called a *MUX*) is an *n*-way digital switch that switches one of *n* inputs to the output. A 2-input multiplexer is shown in Fig. 5.1. The switch is controlled by the single control line *s*. This bit selects one of the two inputs to be "connected" to the output. This means that the logical value of the output *y* will be the same as the logical value of the selected input.

From the truth table in Fig. 5.1 we see that $y = a$ if $s = 0$ and $y = b$ if $s = 1$. The Karnaugh map for the truth table in Fig. 5.1 is shown in Fig. 5.2. We see that the logic equation for *y* is

```
y =  ~s & a | s & b                    (5.1)
```

Note that this logic equation describes the circuit diagram shown in Fig. 5.3.

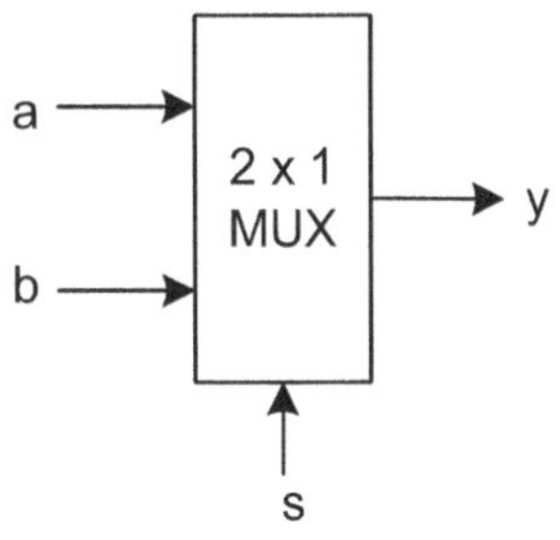

s	a	b	y
0	0	0	0
0	0	1	0
0	1	0	1
0	1	1	1
1	0	0	0
1	0	1	1
1	1	0	0
1	1	1	1

Figure 5.1 A 2-to-1 multiplexer

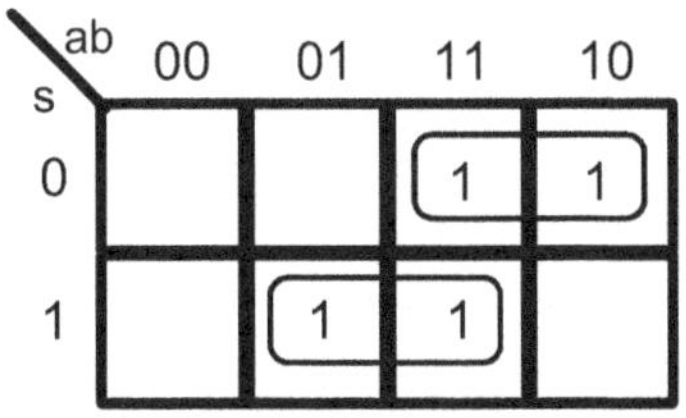

y = ~s & a | s & b

Figure 5.2
K-map for a 2-to-1 multiplexer

Use the BDE to create the block diagram *mux21.bde* shown in Fig. 5.3 that implements logic equation (5.1). Compiling *mux21.bde* will generate a VHDL file, *mux21.vhd*, that is equivalent to Listing 5.1. A simulation of *mux21.bde* is shown in Fig. 5.4. Note in the simulation that $y = a$ if $s = 0$ and $y = b$ if $s = 1$.

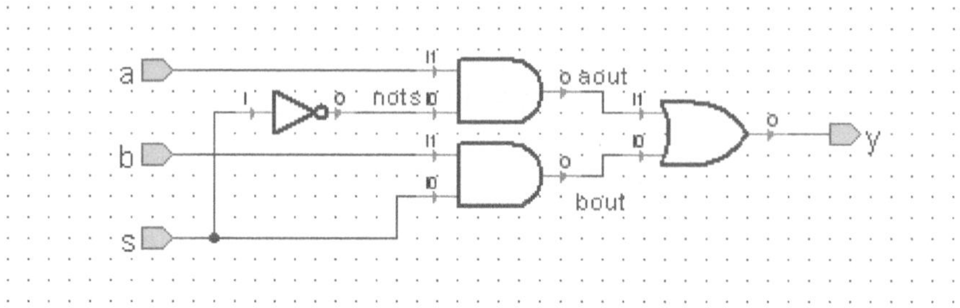

Figure 5.3 Block diagram for a 2-to-1 multiplexer, *mux21.bde*

Listing 5.1 Example5a.vhd

```
-- Example 5a: 2-to-1 MUX using logic equations
library IEEE;
use IEEE.std_logic_1164.all;

entity mux21 is
  port(
        a : in STD_LOGIC;
        b : in STD_LOGIC;
        s : in STD_LOGIC;
        y : out STD_LOGIC
  );
end mux21;

architecture mux21 of mux21 is
signal aout : STD_LOGIC;
signal bout : STD_LOGIC;
signal nots : STD_LOGIC;

begin
      aout <= nots and a;
      bout <= s and b;
      nots <= not(s);
      y <= bout or aout;
end mux21;
```

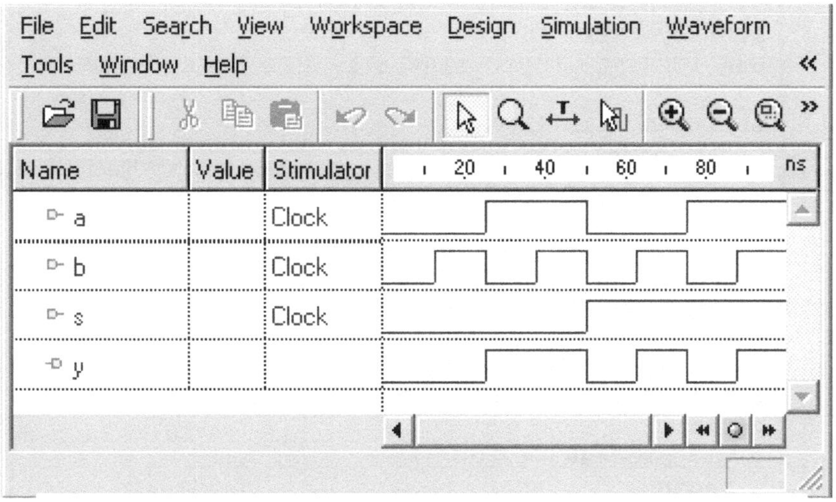

Figure 5.4 Simulation of the 2-to-1 MUX in Fig. 5.3

5.2 The VHDL *if* statement

The behavior of the 2 x 1 multiplexer shown in Fig. 5.1 can be described by the VHDL statements

```
if s = '0' then
   y <= a;
else
   y <= b;
```

We saw that the assignment statements in VHDL using the assignment operator <= are concurrent and execute in parallel. On the other hand the *if* statement is an example of a *procedural*, or *sequential*, statement. Procedural statements must be contained within a *process* and are executed in the order that they appear in the code. Thus, the VHDL *if* statement must be contained in a *process* as shown in Listing 5.2.

The *process* begins with the statement

```
<label>: process(<sensitivity_list>)
```

where the *sensitivity list* contains a list of all signals that will affect the outputs generated by the *process* block and the *label* is an arbitrary name of your choice following typical variable naming conventions. In Listing 5.2 the sensitivity list contains the inputs *a*, *b*, and *s*, so that a change in any of these three inputs will affect the output *y*. If you do not include a signal in the sensitivity list then the circuit that is generated may not be the one that you want. This is a common error that is sometimes hard to detect. The VHDL code in Listing 5.2 will be compiled to produce the logic circuit shown in Fig. 5.3. A simulation of the VHDL code in Listing 5.2 will produce the same waveform as shown in Fig. 5.4.

Listing 5.2 Example4b.vhd

```
-- Example 4b: 2-to-1 MUX using if statement
library IEEE;
use IEEE.STD_LOGIC_1164.all;

entity mux21b is
      port(
            a : in STD_LOGIC;
            b : in STD_LOGIC;
            s : in STD_LOGIC;
            y : out STD_LOGIC
          );
end mux21b;

architecture mux21b of mux21b is
begin
  p1: process (a, b, s)
  begin
     if s = '0' then
       y <= a;
     else
       y <= b;
     end if;
  end process;
end mux21b;
```

Create a top-level design called *mux21_top.bde* that connects *a* and *b* to the rightmost two slide switches, connects *s* to *btn*(0), and connects the output *y* to *ld*(0). Implement your design and download it to the FPGA board. Test the operation of the multiplexer by changing the position of the toggle switches and pressing pushbutton *btn*(0).

Example 6

Quad 2-to-1 Multiplexer

In this example we will show how to design a quad 2-to-1 multiplexer. In Section 6.1 we will make the quad 2-to-1 multiplexer by wiring together four of the 2-to-1 multiplexers that we designed in Example 5. In Section 6.2 we will show how the quad 2-to-1 multiplexer can be designed using a single VHDL *if* statement. Finally, in Section 6.3 we will show how to use a VHDL parameter to define a generic 2-to-1 multiplexer with arbitrary bus sizes.

Prerequisite knowledge:
Example 5 – 2-to-1 Multiplexer

6.1 Generating the Design File *mux42.bde*

By using four instances of the 2-to-1 MUX, *mux21.bde*, that we designed in Example 5, we can design a quad 2-to-1 multiplexer as shown in Fig. 6.1. Use the BDE to create the file *mux24.bde* using Active-HDL. Note that you will need to add the file *mux21.bde* to your project.

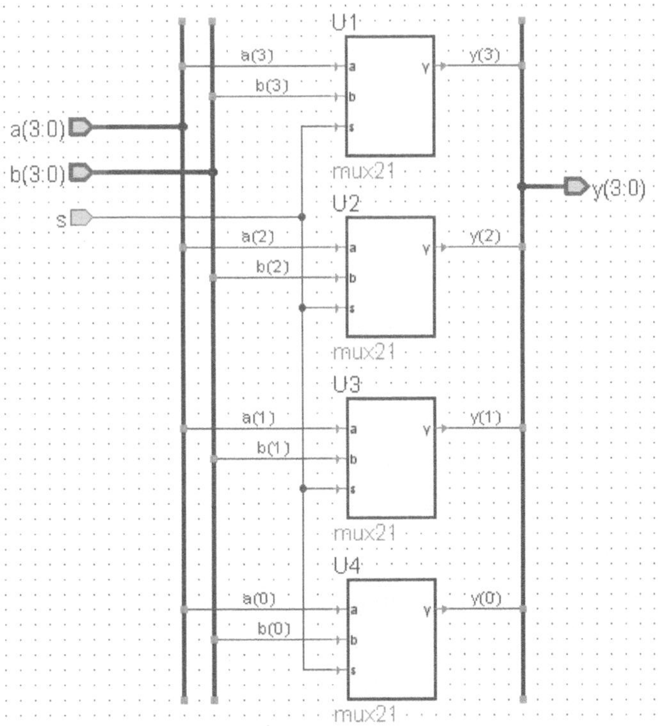

Figure 6.1 The quad 2-to-1 MUX, *mux24.bde*, contains four 2-to-1 MUXs

If you compile the file *mux24.bde* Active-HDL will generate the VHDL program *mux24.vhd* shown in Listing 6.1. A simulation of *mux24.bde* is shown in Fig. 6.2. Note that the output *y*(3:0) will be either *a*(3:0) or *b*(3:0) depending on the value of *s*.

Listing 6.1 Example6a.vhd

```
-- Example 6a: mux24
library IEEE;
use IEEE.std_logic_1164.all;
library EXAMPLE6;

entity mux24 is
  port(
       s : in std_logic;
       a : in STD_LOGIC_VECTOR(3 downto 0);
       b : in STD_LOGIC_VECTOR(3 downto 0);
       y : out STD_LOGIC_VECTOR(3 downto 0)
  );
end mux24;

architecture mux24 of mux24 is
component mux21
  port (
       a : in std_logic;
       b : in std_logic;
       s : in std_logic;
       y : out std_logic
  );
end component;

begin
U1 : mux21
  port map(
       a => a(3), b => b(3), s => s, y => y(3)
  );

U2 : mux21
  port map(
       a => a(2), b => b(2), s => s, y => y(2)
  );

U3 : mux21
  port map(
       a => a(1), b => b(1), s => s, y => y(1)
  );

U4 : mux21
  port map(
       a => a(0), b => b(0), s => s, y => y(0)
  );

end mux24;
```

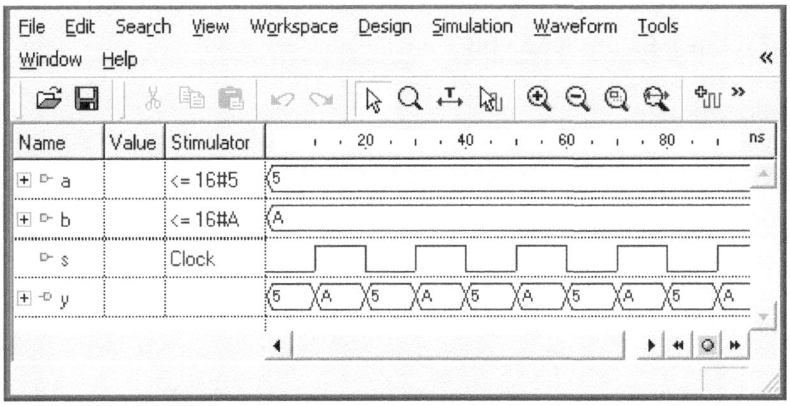

Figure 6.2 Simulation of the quad 2-to-1 MUX in Fig. 6.1

Use the BDE to create the top-level design called *mux21_top.bde* shown in Fig. 6.3. Note that *a*(3:0) are connected to the four leftmost slide switches, *b*(3:0) are connected to the rightmost four slide switches, and *y*(3:0) are connected to the four rightmost LEDs. Also note that *s* is connected to *btn*(0), and the input *btn*(0:0) must be declared as a *std_logic_vector*, even though there is only one element, so that we can use the constraint file *basys2.ucf* or *nexys2.ucf* without change. Implement your design and download it to the FPGA board. Test the operation of the quad 2-to-1 multiplexer by setting the switch values and pressing pushbutton *btn*(0).

If you compile the file *mux24_top.bde* Active-HDL will generate the VHDL program *mux24_top.vhd* shown in Listing 6.2. A simulation of *mux24_top.bde* is shown in Fig. 6.4.

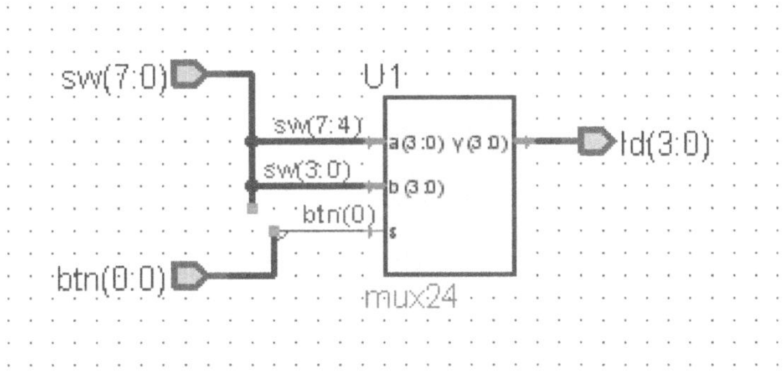

Figure 6.3 Top-level design for testing the quad 2-to-1 MUX

Listing 6.2 Example6b.vhd

```
-- Example 6b: mux24_top
library IEEE;
use IEEE.std_logic_1164.all;
library EXAMPLE6;

entity mux24_top is
  port(
       btn : in STD_LOGIC_VECTOR(0 downto 0);
       sw : in std_logic_vector(7 downto 0);
       ld : out std_logic_vector(3 downto 0)
  );
end mux24_top;
```

Listing 6.2 (cont.) Example6b.vhd

```
architecture mux24_top of mux24_top is
component mux24
  port (
       a : in std_logic_vector(3 downto 0);
       b : in std_logic_vector(3 downto 0);
       s : in std_logic;
       y : out std_logic_vector(3 downto 0)
  );
end component;

begin
U1 : mux24
  port map(
       a(0) => sw(4),
       a(1) => sw(5),
       a(2) => sw(6),
       a(3) => sw(7),
       b(0) => sw(0),
       b(1) => sw(1),
       b(2) => sw(2),
       b(3) => sw(3),
       s => btn(0),
       y => ld
  );
end mux24_top;
```

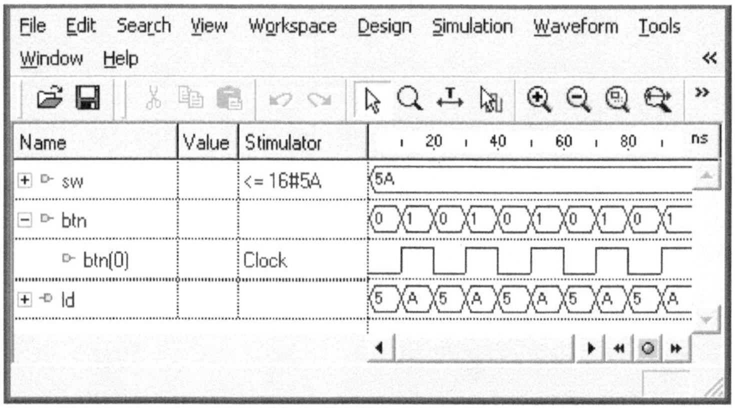

Figure 6.4 Simulation of *mux24_top.bde* in Fig. 6.1

6.2 A Quad 2-to-1 Multiplexer Using an *if* Statement

In Listing 5.2 of Example 5 we used a VHDL *if* statement to implement a 2-to-1 MUX. Listing 6.3 is a direct extension of Listing 5.2 where now the inputs and outputs are 4-bit values rather that a single bit. The VHDL program shown in Listing 6.3 will produce the same simulation as shown in Fig. 6.2. The module *mux24b* defined by the VHDL program in Listing 6.3 could be used in place of the *mux24* module in the top-level design in Fig. 6.3

Listing 6.3 mux24b.vhd

```
--Example 6c: Quad 2-to-1 MUX using if statement
library IEEE;
use IEEE.STD_LOGIC_1164.all;

entity mux24b is
       port(
              a : in STD_LOGIC_VECTOR(3 downto 0);
              b : in STD_LOGIC_VECTOR(3 downto 0);
              s : in STD_LOGIC;
              y : out STD_LOGIC_VECTOR(3 downto 0)
            );
end mux24b;

architecture mux24b of mux24b is
signal s4: STD_LOGIC_VECTOR(3 downto 0);
begin
  p1: process (a, b, s)
  begin
      if s = '0' then
        y <= a;
      else
        y <= b;
      end if;
  end process;
end mux24b;
```

6.3 Generic Multiplexers: Parameters

We can use the VHDL *generic* statement to design a generic 2-to-1 multiplexer with input and output bus widths of arbitrary size. Listing 6.4 shows a VHDL program for a generic 2-to-1 MUX.

Note the use of the *generic* statement that defines the bus width *N* to have a default value of 4. This value can be overridden when the multiplexer is instantiated as shown in Listing 6.5 for an 8-line 2-to-1 multiplexer called *M8*. The parameter override clause is automatically included in the *port map* statement when you copy it in Active-HDL as shown in Listing 6.5. We will always use upper-case names for parameters. The simulation of Listing 6.5 is shown in Fig. 6.5.

If you compile the VHDL program *mux2g.vhd* shown in Listing 6.4 it will generate a block diagram for this module when you go to BDE. If you right-click on the symbol for *mux2g* and select *Properties*, you can change the default value of the parameter *N* by selecting the *Parameters* tab and entering an actual value for *N*.

Listing 6.4 mux2g.vhd

```
-- Example 6d: Generic 2-to-1 MUX using a parameter
library IEEE;
use IEEE.STD_LOGIC_1164.all;

entity mux2g is
      generic(N:integer := 4);
       port(
             a : in STD_LOGIC_VECTOR(N-1 downto 0);
             b : in STD_LOGIC_VECTOR(N-1 downto 0);
             s : in STD_LOGIC;
             y : out STD_LOGIC_VECTOR(N-1 downto 0)
           );
end mux2g;

architecture mux2g of mux2g is
begin
  p1: process (a, b, s)
  begin
      if s = '0' then
        y <= a;
      else
        y <= b;
      end if;
  end process;

end mux2g;
```

Listing 6.5 mux28.vhd

```
--Example 6e: 8-line 2-to-1 MUX using a parameter
library IEEE;
use IEEE.STD_LOGIC_1164.all;

entity mux28 is
       port(
             a : in STD_LOGIC_VECTOR(7 downto 0);
             b : in STD_LOGIC_VECTOR(7 downto 0);
             s : in STD_LOGIC;
             y : out STD_LOGIC_VECTOR(7 downto 0)
           );
end mux28;

architecture mux28 of mux28 is
component mux2g is
      generic(N: positive := 4);
       port(
             a : in STD_LOGIC_VECTOR(N-1 downto 0);
             b : in STD_LOGIC_VECTOR(N-1 downto 0);
             s : in STD_LOGIC;
             y : out STD_LOGIC_VECTOR(N-1 downto 0)
           );
end component;
```

Listing 6.5 (cont.) mux28.vhd

```
begin

M8: mux2g generic map(N => 8) port map
    (a => a,
     b => b,
     s => s,
     y => y
    );

end mux28;
```

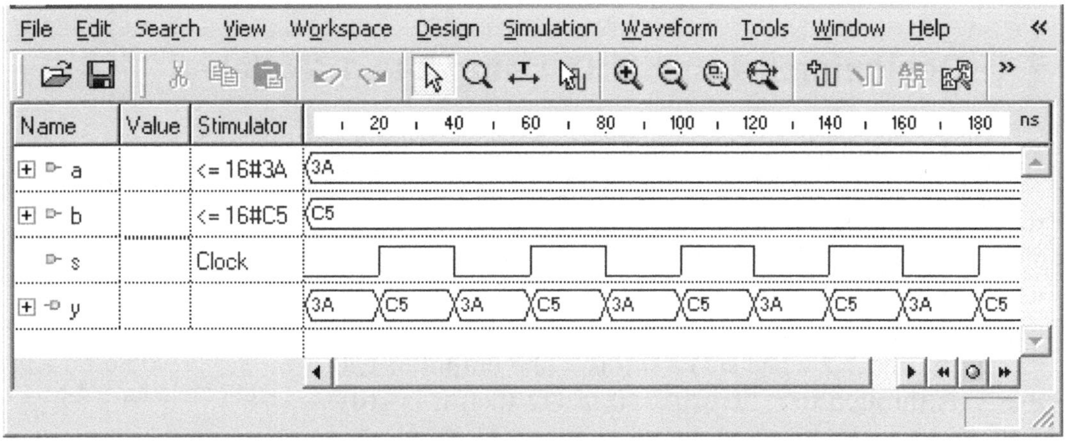

Figure 6.5 Simulation result from the VHDL program in Listing 6.5

Example 7

4-to-1 Multiplexer

In this example we will show how to design a 4-to-1 multiplexer. In Section 7.1 we will make a 4-to-1 multiplexer by wiring together three of the 2-to-1 multiplexers that we designed in Example 5. In Section 7.2 we will derive the logic equation for a 4-to-1 MUX. In Section 7.3 we will show how a 4-to-1 multiplexer can be designed using a single VHDL *case* statement and in Section 7.4 we design a quad 4-to-1 multiplexer.

Prerequisite knowledge:
Example 5 – 2-to-1 Multiplexer

7.1 Designing a 4-to-1 MUX Using 2-to-1 Modules

A 4-to-1 multiplexer has the truth table shown in Fig. 7.1 By using three instances of the 2-to-1 MUX, *mux21.bde*, that we designed in Example 5, we can design a 4-to-1 multiplexer as shown in Fig. 7.2. Use the BDE to create the file *mux41.bde* using Active-HDL. Note that you will need to add the file *mux21.bde* to your project.

s1	s0	z
0	0	c0
0	1	c1
1	0	c2
1	1	c3

Figure 7.1
Truth table for a 4-to-1 MUX

In Fig. 7.2 when *s*(1) = 0 it is *v*, the output of U2 that gets through to *z*. If *s*(0) = 0 in U2 then it is *c*(0) that gets through to *v* and therefore to *z*. If *s*(0) = 1 in U2 then it is *c*(1) that gets through to *v* and therefore to *z*.

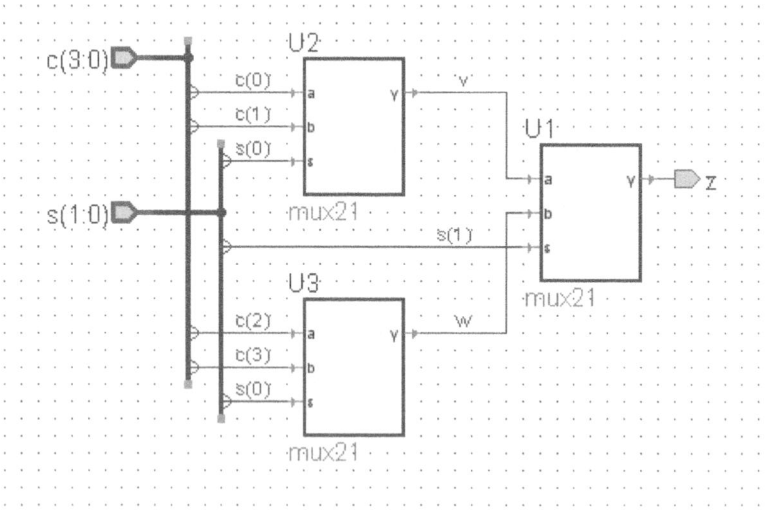

Figure 7.2 The 4-to-1 MUX, *mux41.bde*, contains four 2-to-1 MUXs

If, on the other hand, $s(1) = 1$ in U1 then it is *w*, the output of U3 that gets through to *z*. If $s(0) = 0$ in U3 then it is $c(2)$ that gets through to *w* and therefore to *z*. If $s(0) = 1$ in U3 then it is $c(3)$ that gets through to *w* and therefore to *z*. Thus you can see that the circuit in Fig. 7.2 will implement the truth table in Fig. 7.1.

When you compile the file *mux41.bde* Active-HDL will generate the VHDL program *mux41.v* shown in Listing 7.1. A simulation of *mux41.bde* is shown in Fig. 7.3. Note that the output *z* will be one of the four inputs $c(3{:}0)$ depending on the value of $s(1{:}0)$.

Listing 7.1 mux41.vhd

```
-- Example 7a: 4-to-1 MUX using module instantiation
library IEEE;
use IEEE.std_logic_1164.all;
library EXAMPLE7;

entity mux41a is
  port(
          c : in STD_LOGIC_VECTOR(3 downto 0);
          s : in STD_LOGIC_VECTOR(1 downto 0);
          z : out std_logic
  );
end mux41a;

architecture mux41a of mux41a is
component mux21
  port (
          a : in std_logic;
          b : in std_logic;
          s : in std_logic;
          y : out std_logic
  );
end component;

signal v : std_logic;
signal w : std_logic;

begin
U1  : mux21
  port map(
          a => v, b => w, s => s(1), y => z
  );

U2  : mux21
  port map(
          a => c(0), b => c(1), s => s(0), y => v
  );

U3  : mux21
  port map(
          a => c(2), b => c(3), s => s(0), y => w
  );

end mux41a;
```

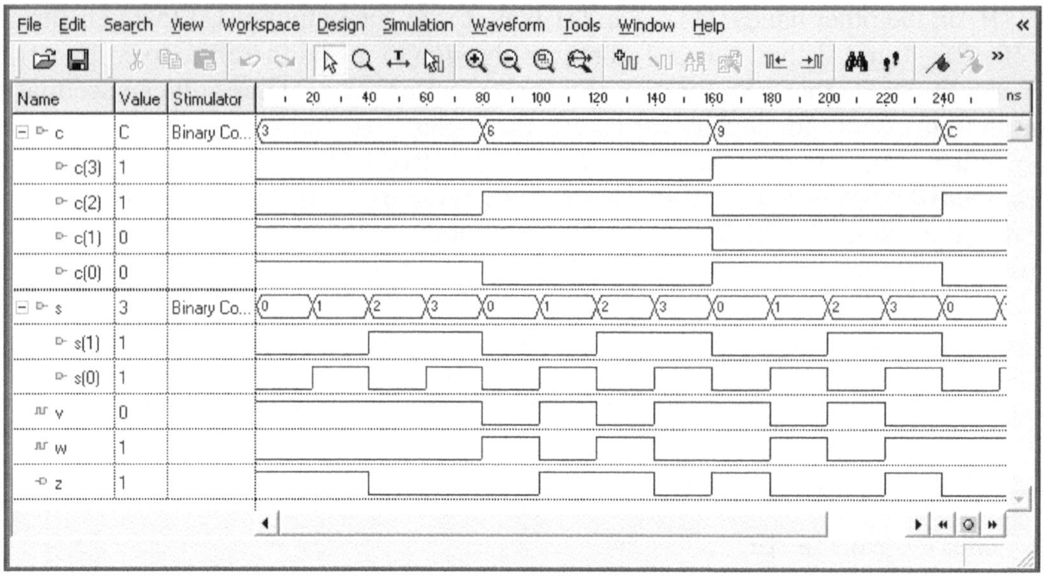

Figure 7.3 Simulation of the VHDL program in Listing 7.1

If you were going to create this top-level design using HDE instead of BDE you would begin by defining the inputs *c*(3:0) and *s*(1:0) and the output *z* and the two signals *v* and *w*. You would then "wire" the three components together using the three *port map statements* shown in Listing 7.1.

The easiest way to generate this *port map* statement is to first compile the file *mux21.vhd* from Example 5 using Active-HDL, expand the library icon (click the plus sign), right click on *mux21*, and select *Copy VHDL Instantiation* as shown in Fig. 7.4. Paste this into your top-level *mux41.vhd* file.

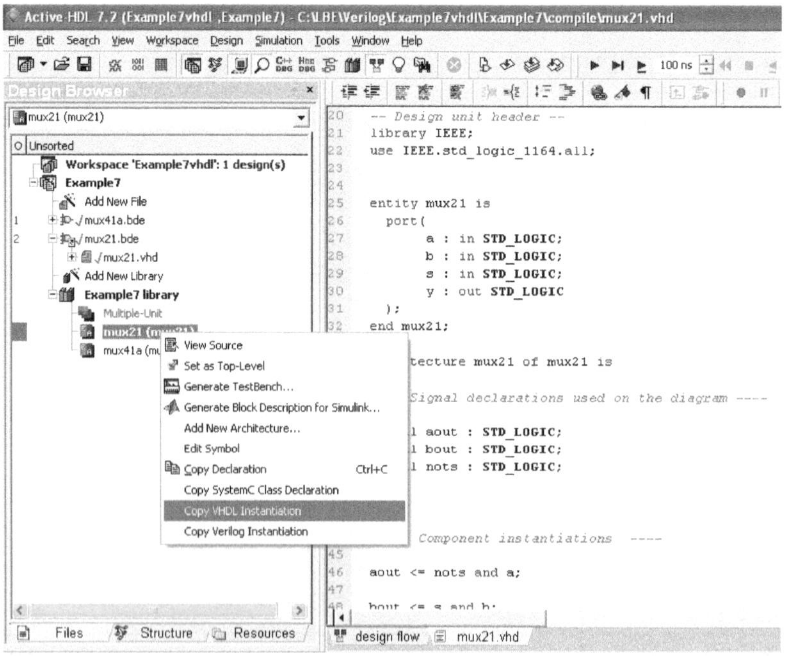

Figure 7.4 Generating a module instantiation prototype

At this point you would have the statement

```
Label1 : mux21
        port map(
                a => a,
                b => b,
                s => s,
                y => y
        );
```

Make three copies of this prototype and change the name of *Label1* to U1, U2, and U3 in the three statements. Now you just "wire up" each input and output variable by changing the values in the parentheses to the signal that it is connected to. For example, the mux U1 input *a* is connected to the wire *v* so we would write `a => v`. In a similar way the mux input *b* is connected to wire *w* and the mux input *s* is connected to input *s*(1). The mux output *y* is connected to the output *z* in Fig. 7.2. Thus, the final version of this *port map* statement would be

```
U1 : mux21
        port map(
                a => v,
                b => w,
                s => s(1),
                y => z
        );
```

The other two modules, U2 and U3, are "wired up" using similar *port map* statements.

7.2 The Logic Equation for a 4-to-1 MUX

The 4-to-1 MUX designed in Fig. 7.2 can be represented by the logic symbol shown in Fig. 7.5. This multiplexer acts like a digital switch in which one of the inputs *c*(3:0) gets connected to the output *z*. The switch is controlled by the two control lines *s*(1:0). The two bits on these control lines select one of the four inputs to be "connected" to the output. Note that we constructed this 4-to-1 multiplexer using three 2-to-1 multiplexers in a tree fashion as shown in Fig. 7.2.

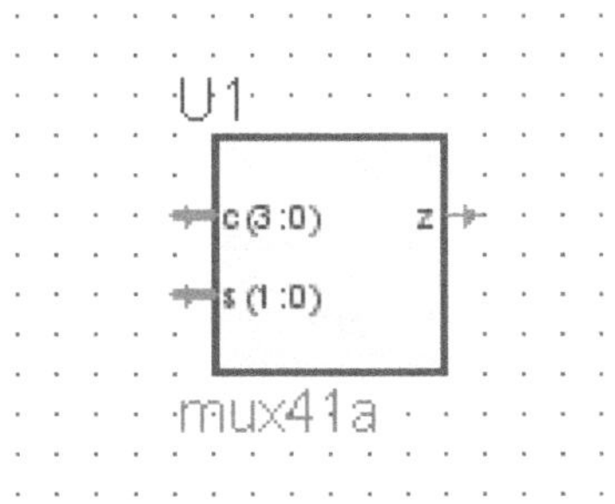

Figure 7.5 A 4-to-1 multiplexer

Recall from Eq. (5.1) in Example 5 that the logic equation for a 2-to-1 MUX is given by

```
y =  ~s & a | s & b                (7.1)
```

Applying this equation to the three 2-to-1 MUXs in Fig. 7.2 we can write the equations for that 4 x 1 MUX as follows.

```
v = ~s0 & c0 | s0 & c1

w = ~s0 & c2 | s0 & c3

z = ~s1 & v | s1 & w

z = ~s1 & (~s0 & c0 | s0 & c1) | s1 & (~s0 & c2 | s0 & c3)
```

or

```
z = ~s1 & ~s0 & c0
  | ~s1 &  s0 & c1                 (7.2)
  |  s1 & ~s0 & c2
  |  s1 &  s0 & c3
```

Equation (7.2) for *z* also follows from the truth table in Fig. 7.1. Note that the tree structure in Fig. 7.2 can be expanded to implement an 8-to-1 multiplexer and a 16-to-1 multiplexer.

A VHDL program that implements a 4-to-1 MUX using the logic equation (7.2) is given in Listing 7.2. A simulation of this program will produce the same result as in Fig. 7.3 (without the wire signals *v* and *w*).

Listing 7.2 mux41b.vhd

```
-- Example 7b: 4-to-1 MUX using logic equation
library IEEE;
use IEEE.STD_LOGIC_1164.all;

entity mux41b is
       port(
              c : in STD_LOGIC_VECTOR(3 downto 0);
              s : in STD_LOGIC_VECTOR(1 downto 0);
              z : out STD_LOGIC
            );
end mux41b;

architecture mux41b of mux41b is

begin
   z <= (not s(1) and not s(0) and c(0))
     or (not s(1) and     s(0) and c(1))
     or (    s(1) and not s(0) and c(2))
     or (    s(1) and     s(0) and c(3));
end mux41b;
```

7.3 4-to-1 Multiplexer: *case* Statement

The same 4-to-1 multiplexer defined by the VHDL program in Listing 7.2 can be implemented using a VHDL *case* statement. The VHDL program shown in Listing 7.3 does this. The *case* statement in Listing 7.3 directly implements the definition of a 4-to-1 MUX given by the truth table in Fig. 7.1. The *case* statement is an example of a *procedural statement* that must be within a *process*. A typical line in the *case* statement, such as

```
when "10" => z <= c(2);
```

will assign the value of *c*(2) to the output *z* when the input value *s*(1:0) is equal to 2 (binary 10).

In the *case* statement the value following the *when* statement represents the value of the *case* parameter, in this case the 2-bit input *s*. These values are the same as the *case* parameter type by default, in this case STD_LOGIC_VECTOR(1:0). If you want to write a hex value you precede the number with *X* as in *X"A"* which is a hex value *A*. However, hex values are in multiples of 4 bits, therefore *X"A"* represents a binary *1010*. Since *s* is only 2 bits, we can't use the hex notation because the bus sizes would not match.

Listing 7.3 mux41c.vhd

```
-- Example 7c:  4-to-1 MUX using case statement
library IEEE;
use IEEE.STD_LOGIC_1164.all;

entity mux41c is
       port(
              c : in STD_LOGIC_VECTOR(3 downto 0);
              s : in STD_LOGIC_VECTOR(1 downto 0);
              z : out STD_LOGIC
           );
end mux41c;

architecture mux41c of mux41c is

begin
   p1: process(c, s)
   begin
      case s is
        when "00" => z <= c(0);
        when "01" => z <= c(1);
        when "10" => z <= c(2);
        when "11" => z <= c(3);
        when others => z <= c(0);
      end case;
   end process;
end mux41c;
```

All *case* statements should include a *when others* line as shown in Listing 7.3. This is because all cases need to be covered and while it looks as if we covered all cases in Listing 7.3, recall that VHDL actually defines *nine* possible values for each bit of type STD_LOGIC_VECTOR.

A simulation of the program in Listing 7.3 will produce the same result as in Fig. 7.3 (without the wire signals *v* and *w*).

7.4 A Quad 4-to-1 Multiplexer

To make a quad 4-to-1 multiplexer we could combine four 4-to-1 MUXs as we did for a quad 2-to-1 multiplexer module in Fig. 6.1 of Example 6. However, it will be easier to modify the *case* statement program in Listing 7.3 to make a quad 4-to-1 MUX. Because we will use it in Example 10 we will define a single 16-bit input *x*(15:0) and we will multiplex the four hex digits making up this 16-bit value.

Listing 7.4 is a VHDL program for this quad 4-to-1 multiplexer. Note that the four hex digits making up the 16-bit value of *x*(15:0) are multiplexed to the output *z*(3:0) depending of the value of the control signal *s*(1:0). A simulation of this quad 4-to-1 multiplexer is shown in Fig. 7.6 and its BDE symbol is shown in Fig. 7.7.

Listing 7.4 mux44.vhd

```
-- Example 7d: quad 4-to-1 MUX
library IEEE;
use IEEE.STD_LOGIC_1164.all;

entity mux44 is
       port(
             x : in STD_LOGIC_VECTOR(15 downto 0);
             s : in STD_LOGIC_VECTOR(1 downto 0);
             z : out STD_LOGIC_VECTOR(3 downto 0)
           );
end mux44;

architecture mux44 of mux44 is

begin
   p1: process(x, s)
   begin
      case s is
        when "00" => z <= x(3 downto 0);
        when "01" => z <= x(7 downto 4);
        when "10" => z <= x(11 downto 8);
        when "11" => z <= x(15 downto 12);
        when others => z <= x(3 downto 0);
      end case;
   end process;

end mux44;
```

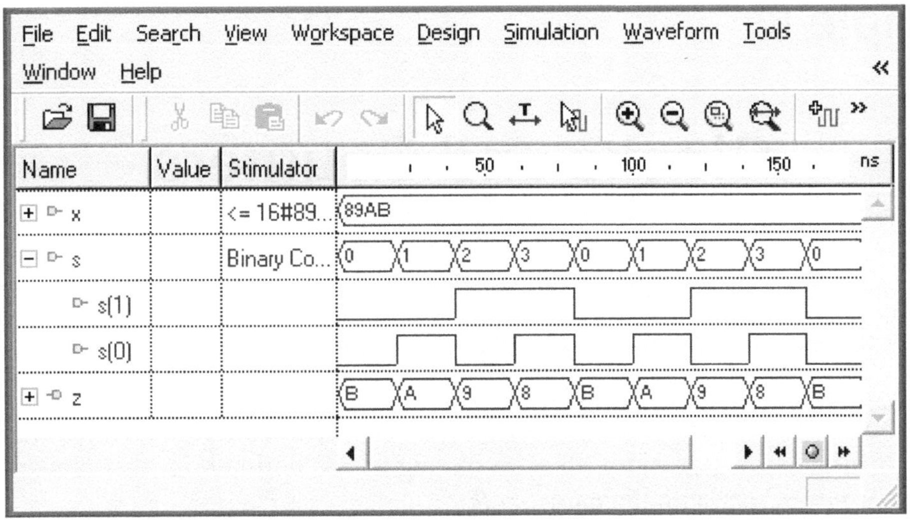

Figure 7.6 Simulation of the quad 4-to-1 MUX in Listing 7.4

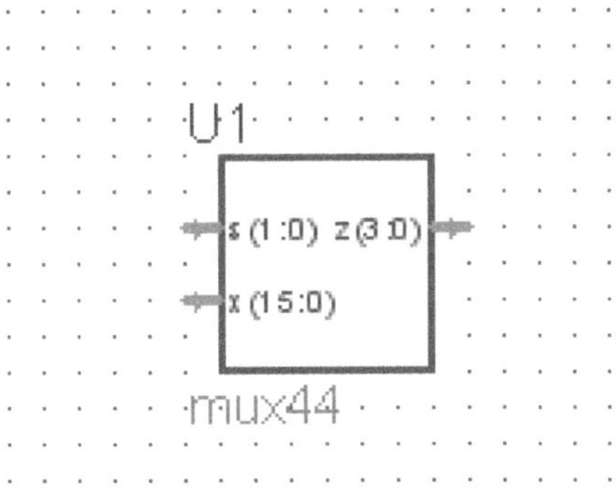

Figure 7.7 A quad 4-to-1 multiplexer

Example 8

Clocks and Counters

The Nexys-2 board has an onboard 50 MHz clock. The BASYS board has a jumper that allows you to set the clock to 100 MHz, 50 MHz, or 25 MHz. All of the examples in this book will assume an input clock frequency of 50 MHz. If you are using the BASYS board you should remove the clock jumper, which will set the clock frequency to 50 MHz. This 50 MHz clock signal is a square wave with a period of 20 ns. The FPGA pin associated with this clock signal is defined in the constraints file *basys2.ucf* or *nexys2.ucf* with the name *mclk*.

In this example we will show how to design an *N*-bit counter in VHDL and how to use a counter to generate clock signals of lower frequencies.

Prerequisite knowledge:

Appendix A – Use of Aldec Active-HDL

8.1 *N*-Bit Counter

The BDE symbol for an *N*-bit counter is shown in Fig. 8.1. If the input *clr* = 1 then all *N* of the outputs *q*(*i*) are cleared to zero asynchronously, i.e., regardless of the value of the input *clk*. If *clr* = 0, then on the next rising edge of the clock input *clk* the *N*-bit binary output *q*(*N*-1:0) will be incremented by 1. That is, on the rising edge of the clock the *N*-bit binary output *q*(*N*-1:0) will count from 0 to *N*-1 and then wrap around to 0.

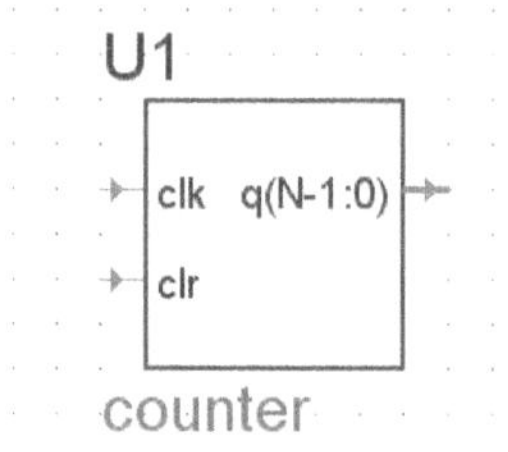

Figure 8.1 An *N*-bit counter

The VHDL program shown in Listing 8.1 was used to generate the symbol shown in Fig. 8.1. Note that the sensitivity list of the process contains the signals *clk* and *clr*. This means that the *if* statement within the *process* will execute whenever either *clr* or *clk* goes high. If *clr* goes high then the output *q*(*N*-1:0) will go to zero. The statement

```
count <= (others => '0');
```

sets all bits of *count*(*N*-1:0) to zero.

The phrase

```
clk'event and clk = '1'
```

in the *elsif* clause in Listing 8.1 means that there was an event on the signal *clk*, i.e., it changed value and it ended up at 1. That is, there was a rising edge of the clock. Thus, if *clr* = 0 and there is a rising edge of the clock signal *clk* then the output *q*(*N*-1:0) will be incremented by 1. Note that *count*(*N*-1:0) is defined to be a *signal* in Listing 8.1. This is necessary because the output *q* can not be read and therefore you can *not* use a statement such as

```
q <= q + 1;
```

in Listing 8.1. Rather you must increment the signal *count*(*N*-1:0) within the process in Listing 8.1 and then assign the output *q* to *count* outside the process.

Listing 8.1 counter.vhd

```
-- Example 8a: N-bit counter
library IEEE;
use IEEE.STD_LOGIC_1164.all;
use IEEE.STD_LOGIC_unsigned.all;

entity counter is
        generic(N : integer := 8);
        port(
                clr : in STD_LOGIC;
                clk : in STD_LOGIC;
                q : out STD_LOGIC_VECTOR(N-1 downto 0)
             );
end counter;

architecture counter of counter is
signal count: STD_LOGIC_VECTOR(N-1 downto 0);
begin
        process(clk, clr)
        begin
                if clr = '1' then
                        count <= (others => '0');
                elsif clk'event and clk = '1' then
                        count <= count + 1;
                end if;
        end process;

        q <= count;
end counter;
```

The default value of the parameter *N* in Listing 8.1 is 4. A simulation of this 4-bit counter is shown in Fig. 8.2. Note that this counter counts from 0 to F and then wraps around to 0. To instantiate an 8-bit counter from Listing 8.1 that would count from 0 – 255 (or 00 – FF hex) you would use an instantiation statement something like

```
Cnt16 : counter generic map(
                N => 16
        )
        port map(
                clr => clr, clk => clk, q => q
        );
```

You can also set the value of the parameter *N* from the block diagram editor (BDE) by right-clicking on the symbol in Fig. 8.1 and selecting *Properties* and then the *Parameters* tab. Note in Listing 8.1 that we have included the additional *use* statement

```
use IEEE.STD_LOGIC_unsigned.all;
```

This statement will include the library file *unsigned.vhd* in the project. This is required in order to use the + sign to implement the counter by adding 1 to the signal *count*.

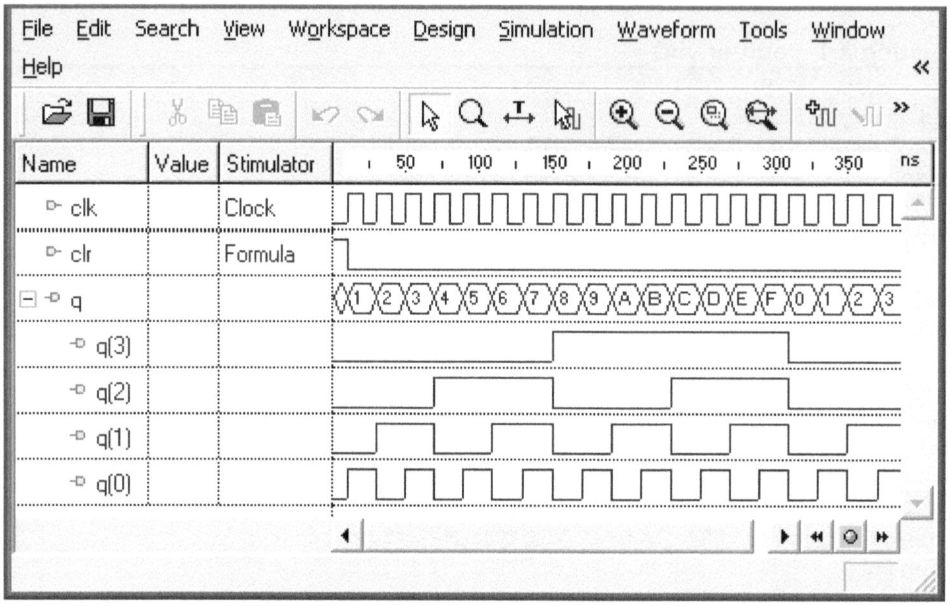

Figure 8.2 Simulation of a 4-bit counter using Listing 8.1

In the simulation in Fig. 8.2 note that the output *q*(0) is a square wave at half the frequency of the input *clk*. Similarly, the output *q*(1) is a square wave at half the frequency of the input *q*(0), the output *q*(2) is a square wave at half the frequency of the input *q*(1), and the output *q*(3) is a square wave at half the frequency of the input *q*(2). Note how the binary numbers *q*(3:0) in Fig. 8.2 count from 0000 to 1111.

The simulation shown in Fig. 8.2 shows how we can obtain a lower clock frequency by simply using one of the outputs *q*(*i*). We will use this feature to produce a 24-bit clock divider in the next section.

8.2 Clock Divider

The simulation in Fig. 8.2 shows that the outputs *q*(*i*) of a counter are square waves where the output *q*(0) has a frequency half of the clock frequency, the output *q*(1) has a frequency half of *q*(0), etc. Thus, a counter can be used to divide the frequency *f* of a clock, where the frequency of the output *q*(*i*) is $f_i = f/2^{i+1}$. The frequencies and periods of the outputs of a 24-bit counter driven by a 50 MHz clock are shown in Table 8.1. Note in Table 8.1 that the output *q*(0) has a frequency of 25 MHz, the output *q*(17) has a frequency of 190.73 Hz, and the output *q*(23) has a frequency of 2.98 Hz.

Table 8.1 Clock divide frequencies

q(*i*)	Frequency (Hz)	Period (ms)
i	50000000.00	0.00002
0	25000000.00	0.00004
1	12500000.00	0.00008
2	6250000.00	0.00016
3	3125000.00	0.00032
4	1562500.00	0.00064
5	781250.00	0.00128
6	390625.00	0.00256
7	195312.50	0.00512
8	97656.25	0.01024
9	48828.13	0.02048
10	24414.06	0.04096
11	12207.03	0.08192
12	6103.52	0.16384
13	3051.76	0.32768
14	1525.88	0.65536
15	762.94	1.31072
16	381.47	2.62144
17	190.73	5.24288
18	95.37	10.48576
19	47.68	20.97152
20	23.84	41.94304
21	11.92	83.88608
22	5.96	167.77216
23	2.98	335.54432

The VHDL program shown in Listing 8.2 is a 24-bit counter that has three outputs, a 25 MHz clock (*clk*25), a 190 Hz clock (*clk*190), and a 3 Hz clock (*clk*3). You can modify this *clkdiv* module to produce any output frequency given in Table 8.1. We will use such a clock divider module in many of our top-level designs.

Listing 8.2 clkdiv.vhd

```
-- Example 8b: clock divider
library IEEE;
use IEEE.STD_LOGIC_1164.all;
use IEEE.STD_LOGIC_unsigned.all;

entity clkdiv is
       port(
             mclk : in STD_LOGIC;
             clr : in STD_LOGIC;
             clk190 : out STD_LOGIC;
             clk48 : out STD_LOGIC
            );
end clkdiv;
```

Listing 8.2 (cont) clkdiv.vhd

```
architecture clkdiv of clkdiv is
signal q:STD_LOGIC_VECTOR(23 downto 0);
begin
  -- clock divider
  process(mclk, clr)
  begin
    if clr = '1' then
       q <= X"000000";
    elsif mclk'event and mclk = '1' then
       q <= q + 1;
    end if;
  end process;

  clk48 <= q(20);          -- 48 Hz
  clk190 <= q(18);         -- 190 Hz
end clkdiv;
```

Note in Listing 8.2 that we define the internal signal *q*(23:0). The BDE symbol generated by compiling Listing 8.2 is shown in Fig. 8.3. You can edit either Listing 8.2 or the block diagram shown in Fig. 8.3 to bring out only the clock frequencies you need in a particular design. For example, the top-level design shown in Fig. 8.4 will cause the eight LEDs on the FPGA board to count in binary at a rate of about three counts per second. The corresponding top-level VHDL program is shown in Listing 8.3.

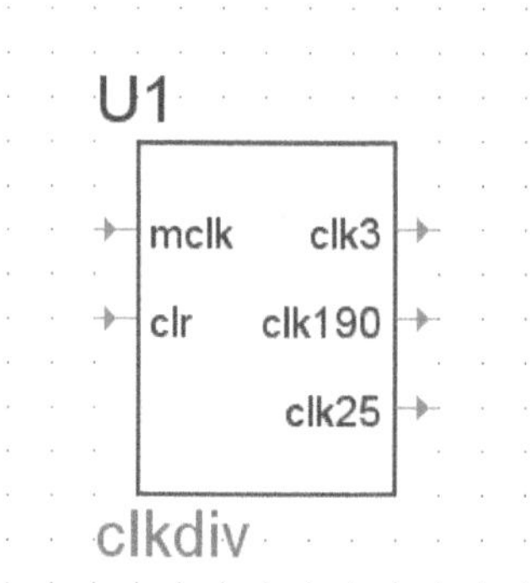

Figure 8.3 A clock divider

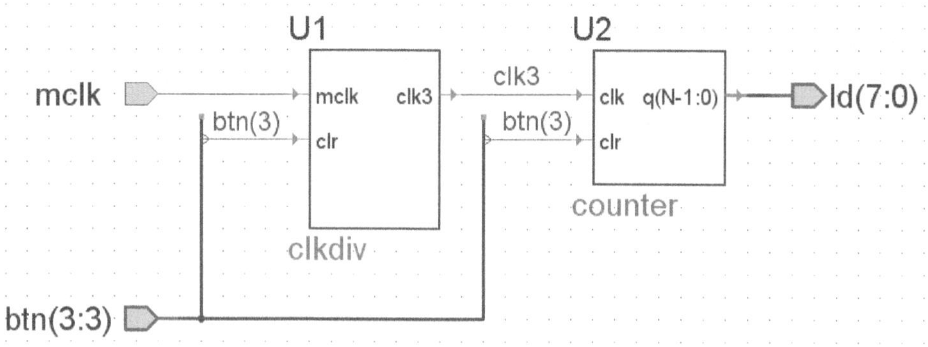

Figure 8.4 Counting in binary on the eight LEDs

Listing 8.3 count8_top.v

```
-- Example 8c: count8_top
library IEEE;
use IEEE.std_logic_1164.all;
library EXAMPLE8;

entity count8_top is
  port(
       mclk : in std_logic;
       btn : in STD_LOGIC_VECTOR(3 downto 3);
       ld : out std_logic_vector(7 downto 0)
  );
end count8_top;

architecture count8_top of count8_top is
component clkdiv
  port (
       clr : in std_logic;
       mclk : in std_logic;
       clk3 : out std_logic
  );
end component;

component counter
  generic(
       N : INTEGER := 8
  );
  port (
       clk : in std_logic;
       clr : in std_logic;
       q : out std_logic_vector(N-1 downto 0)
  );
end component;

signal clk3 : std_logic;

begin
U1 : clkdiv
  port map(
       clk3 => clk3, clr => btn(3), mclk => mclk);

U2 : counter
  generic map (
       N => 8)
  port map(
       clk => clk3, clr => btn(3), q => ld( 7 downto 0 ));

end count8_top;
```

Internally, a counter contains a collection of flip-flops. We saw in Fig. 1 of the *Introduction* that each of the four slices in a CLB of a Spartan3E FPGA contains two flip-flops. Such flip-flops are central to the operation of all synchronous sequential circuits in which changes take place on the rising edge of a clock. The examples in the second half of this book will involve sequential circuits beginning with an example of an edge-triggered D flip-flop in Example 16.

Example 9

7-Segment Decoder

In this section we will show how to design a 7-segment decoder using Karnaugh maps and write a VHDL program to implement the resulting logic equations. We will also solve the same problem using a VHDL *case* statement.

Prerequisite knowledge:

Karnaugh maps – Appendix D
case statement – Example 7
LEDs – Example 1

9.1 7-Segment Displays

Seven LEDs can be arranged in a pattern to form different digits as shown in Fig. 9.1. Digital watches use similar 7-segment displays using liquid crystals rather than LEDs. The red digits on digital clocks are LEDs. Seven segment displays come in two flavors: *common anode* and *common cathode.* A common anode 7-segment display has all of the anodes tied together while a common cathode 7-segment display has all the cathodes tied together as shown in Fig. 9.1.

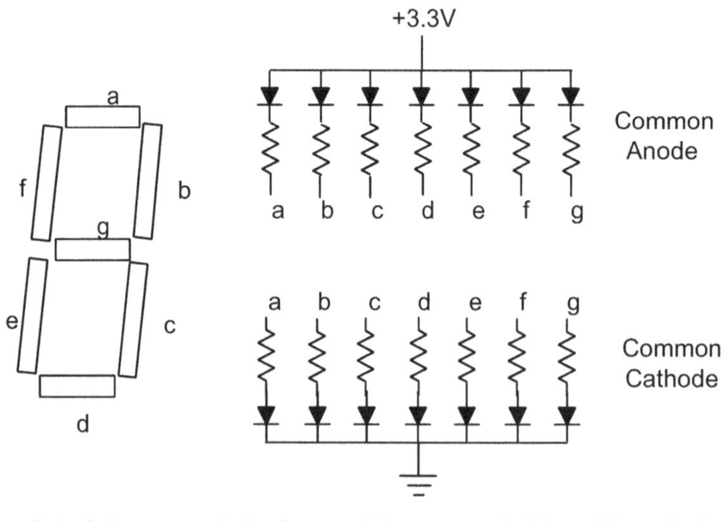

Figure 9.1 A 7-segment display contains seven light emitting diodes (LEDs)

The BASYS and Nexys2 boards have four common-anode 7-segment displays. This means that all the anodes are tied together and connected through a *pnp* transistor to +3.3V. A different FPGA output pin is connected through a 100Ω current-limiting resistor to each of the cathodes, *a* – *g*, plus the decimal point. In the common-anode case, an output 0 will turn on a segment and an output 1 will turn it off. The table shown in

Fig. 9.2 shows output cathode values for each segment *a* – *g* needed to display all hex values from 0 – F.

x	a	b	c	d	e	f	g
0	0	0	0	0	0	0	1
1	1	0	0	1	1	1	1
2	0	0	1	0	0	1	0
3	0	0	0	0	1	1	0
4	1	0	0	1	1	0	0
5	0	1	0	0	1	0	0
6	0	1	0	0	0	0	0
7	0	0	0	1	1	1	1
8	0	0	0	0	0	0	0
9	0	0	0	0	1	0	0
A	0	0	0	1	0	0	0
b	1	1	0	0	0	0	0
C	0	1	1	0	0	0	1
d	1	0	0	0	0	1	0
E	0	1	1	0	0	0	0
F	0	1	1	1	0	0	0

1 = off

0 = on

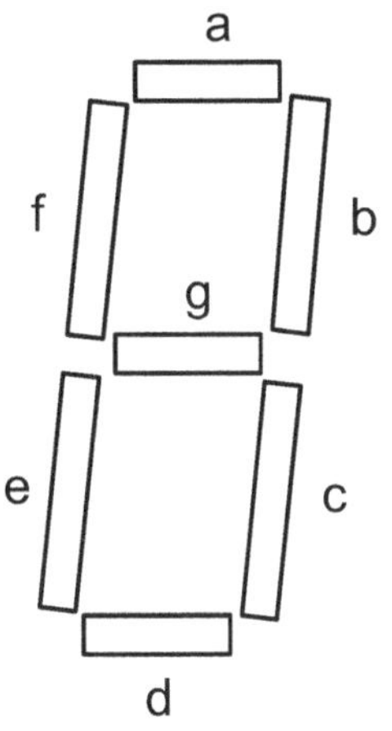

Figure 9.2 Segment values required to display hex digits 0 – F

9.2 7-Segment Decoder: Logic Equations

The problem is to design a *hex to 7-segment decoder*, called *hex7seg*, that is shown in Fig. 9.3. The input is a 4-bit hex number, *x*(3:0), and the outputs are the 7-segment values *a* – *g* given by the truth table in Fig. 9.2. We can make a Karnaugh map for each segment and then write logic equations for the segments *a* – *g*. For example, the K-map for the segment, *e*, is shown in Figure 9.4.

Figure 9.3 A hex to 7-segment decoder

e = ~x3 & x0 | ~x3 & x2 & ~x1 | ~x2 & ~x1 & x0

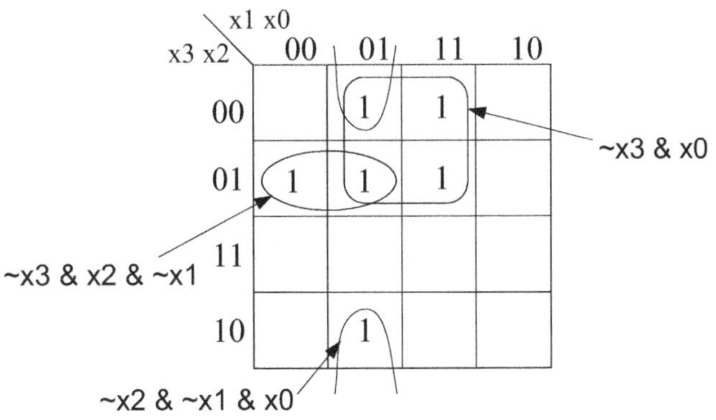

Figure 9.4 K-map for the segment *e* in the 7-segment decoder

You can write the Karnaugh maps for the other six segments and then write the VHDL program for the 7-segment decoder shown in Listing 9.1. A simulation of this program is shown in Fig. 9.5. Note that the simulation agrees with the truth table in Fig. 9.2.

Listing 9.1 hex7seg_le.vhd

```
-- Example 9a: Hex to 7-segment decoder; a-g active low
library IEEE;
use IEEE.STD_LOGIC_1164.all;

entity hex7seg_le is
  port(
        x : in STD_LOGIC_VECTOR(3 downto 0);
        a_to_g : out STD_LOGIC_VECTOR(6 downto 0)
     );
end hex7seg_le;

architecture hex7seg_le of hex7seg_le is
begin
  a_to_g(6) <=  (not x(3) and not x(2) and not x(1) and x(0))--a
             or (not x(3) and x(2) and not x(1) and not x(0))
             or (x(3) and x(2) and not x(1) and x(0))
             or (x(3) and not x(2) and x(1) and x(0));
  a_to_g(5) <=  (x(2) and x(1) and not x(0))                --b
             or (x(3) and x(1) and x(0))
             or (not x(3) and x(2) and not x(1) and x(0))
             or (x(3) and x(2) and not x(1) and not x(0));
  a_to_g(4) <=  (not x(3) and not x(2) and x(1) and not x(0))--c
             or (x(3) and x(2) and x(1))
             or (x(3) and x(2) and not x(0));
  a_to_g(3) <=  (not x(3) and not x(2) and not x(1) and x(0))--d
             or (not x(3) and x(2) and not x(1) and not x(0))
             or (x(3) and not x(2) and x(1) and not x(0))
             or (x(2) and x(1) and x(0));
  a_to_g(2) <=  (not x(3) and x(0))                          --e
             or (not x(3) and x(2) and not x(1))
             or (not x(2) and not x(1) and x(0));
  a_to_g(1) <=  (not x(3) and not x(2) and x(0))             --f
             or (not x(3) and not x(2) and x(1))
             or (not x(3) and x(1) and x(0))
             or (x(3) and x(2) and not x(1) and x(0));
  a_to_g(0) <=  (not x(3) and not x(2) and not x(1))         --g
             or (x(3) and x(2) and not x(1) and not x(0))
             or (not x(3) and x(2) and x(1) and x(0));
end hex7seg_le;
```

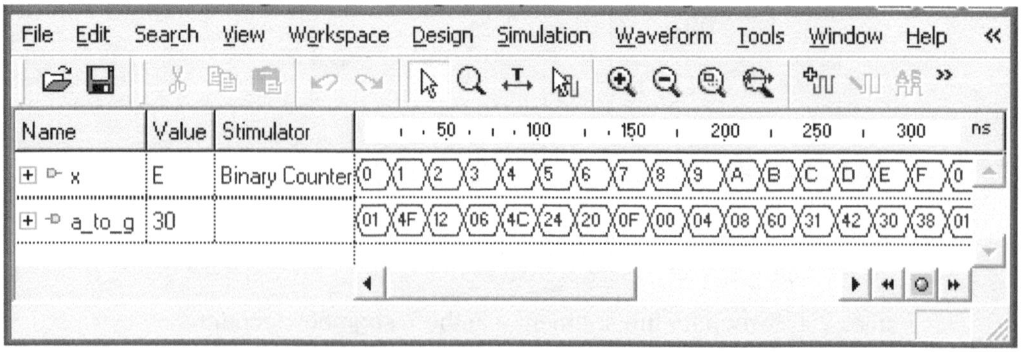

Figure 9.5 Simulation of the VHDL program in Listing 9.1

9.3 7-Segment Decoder: *case* Statement

We can use a VHDL *case* statement to design the same 7-segment decoder that we designed in Section 9.2 using Karnaugh maps. The VHDL program shown in Listing 9.2 is a hex-to-seven-segment decoder that converts a 4-bit input hex digit, *0 – F*, to the appropriate 7-segment codes, *a – g*. The *case* statement in Listing 9.2 directly implements the truth table in Fig. 9.2. Recall that a typical line in the *case* statement, such as

```
when "0011" => a_to_g <= "0000110";   --3
```

will assign the 7-bit binary value, 0000110, to the 7-bit array, `a_to_g`, when the input hex value *x*(3:0) is equal to 3 (0011). In the array `a_to_g` the value `a_to_g`(6) corresponds to segment *a* and the value `a_to_g`(0) corresponds to segment *g*. . Note that in VHDL a hex number is preceded by an *X*.

In the *case* statement the value following the *when* statement in each line represents the value of the *case* parameter, in this case the 4-bit input *x*. The VHDL program in Listing 9.2 shows the implementation of the 7-segment decoder using a *case* statement.

Recall that all *case* statements should include a *when others* line as shown in Listing 9.2. This is because all cases need to be covered and while it looks as if we covered all cases in Listing 9.2, as mentioned previously VHDL actually defines *nine* possible values for each bit of type STD_LOGIC_VECTOR (see Example 1).

A simulation of Listing 9.2 will produce the same results as shown in Fig. 9.5. It should be clear from this example and Example 7 that using the VHDL *case* statement is often easier than solving for the logic equations using Karnaugh maps.

To test the 7-segment displays on the BASYS or Nexys-2 board create a new project and add the files *hex7seg.vhd* from Listing 9.2 and the top-level design *hex7seg_top.vhd* given in Listing 9.3. Each of the four digits on the 7-segment display is enabled by one of the active low signals *an*(3:0) and all digits share the same *a_to_g*(6:0) signals. If *an*(3:0) = 0000 then all digits are enabled and display the same hex digit. This is what we do in Fig. 9.6 and Listing 9.3. Making the output *dp* = 1 will cause the decimal points to be off. You should be able to display all of the hex digits from 0 – F by changing the four right-most switches.

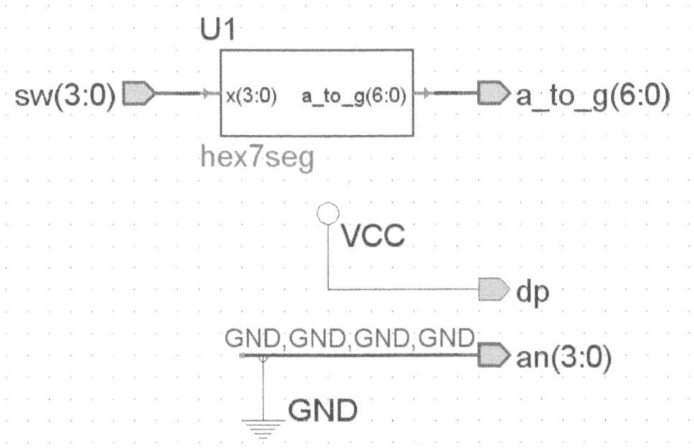

Figure 9.6 Top-level design for testing *hex7seg*

Listing 9.2 hex7seg.vhd

```
-- Example 9b: Hex to 7-segment decoder; a-g active low
library IEEE;
use IEEE.STD_LOGIC_1164.all;

entity hex7seg is
       port(
              x : in STD_LOGIC_VECTOR(3 downto 0);
              a_to_g : out STD_LOGIC_VECTOR(6 downto 0)
            );
end hex7seg;

architecture hex7seg of hex7segbis
begin
    process(x)
    begin
        case x is
             when X"0" => a_to_g <= "0000001";     --0
             when X"1" => a_to_g <= "1001111";     --1
             when X"2" => a_to_g <= "0010010";     --2
             when X"3" => a_to_g <= "0000110";     --3
             when X"4" => a_to_g <= "1001100";     --4
             when X"5" => a_to_g <= "0100100";     --5
             when X"6" => a_to_g <= "0100000";     --6
             when X"7" => a_to_g <= "0001101";     --7
             when X"8" => a_to_g <= "0000000";     --8
             when X"9" => a_to_g <= "0000100";     --9
             when X"A" => a_to_g <= "0001000";     --A
             when X"B" => a_to_g <= "1100000";     --b
             when X"C" => a_to_g <= "0110001";     --C
             when X"D" => a_to_g <= "1000010";     --d
             when X"E" => a_to_g <= "0110000";     --E
             when others => a_to_g <= "0111000";   --F
         end case;
   end process;
end hex7seg;
```

Listing 9.3 hex7seg_top.vhd

```
-- Example 9c: hex7seg_top
library IEEE;
use IEEE.STD_LOGIC_1164.all;

entity hex7seg_top is
       port(
              sw : in STD_LOGIC_VECTOR(3 downto 0);
              a_to_g : out STD_LOGIC_VECTOR(6 downto 0);
              an : out STD_LOGIC_VECTOR(3 downto 0);
              dp : out STD_LOGIC
            );
end seg7test;

architecture hex7seg_top of hex7seg_top is
component hex7seg is
       port(
              x : in STD_LOGIC_VECTOR(3 downto 0);
              a_to_g : out STD_LOGIC_VECTOR(6 downto 0)
            );
end component;

begin
an <= "0000";      --all digits on
dp <= '1';         --dp off

D4: hex7seg port map
    (x => sw,
     a_to_g => a_to_g
    );

end hex7seg_top;
```

Example 10

7-Segment Displays: *x7seg* and *x7segb*

In this example we will show how to display different hex values on the four 7-segment displays.

Prerequisite knowledge:
Karnaugh maps – Appendix D
case statement – Example 7
LEDs – Example 1

10.1 Multiplexing 7-Segment Displays

We saw in Example 9 that the *a_to_g*(6:0) signals go to all of the 7-segment displays and therefore in that example all of the digits displayed the same value. How could we display a 4-digit number such as 1234 that contains different digits? To see how we might do this, consider the BDE circuit shown in Fig. 10.1. Instead of enabling all four digits at once by setting *an*(3:0) = "0000" as we did in Fig. 9.6 we connect *an*(3:0) to the NOT of the four pushbuttons *btn*(3:0). Thus, a digit will only be enabled when the corresponding pushbutton is being pressed.

The quad 4-to-1 multiplexer, *mux44*, from Listing 7.4 is used to display the 16-bit number *x*(15:0) as a 4-digit hex value on the 7-segment displays. When you press *btn*(0) if the control signal *s*(1:0) is 00 then *x*(3:0) becomes the input to the *hex7seg* module and the value of *x*(3:0) will be displayed on digit 0. Similarly if you press *btn*(1) and the control signal *s*(1:0) is 01 then *x*(7:4) becomes the input to the *hex7seg* module and the value of *x*(7:4) will be displayed on digit 1. We can make the value of *s*(1:0) depend on the value of *btn*(3:0) using the truth table in Fig. 10.2. From this truth table we can write the following logic equations for *s*(1) and *s*(0).

```
s(1) <= btn(2) or btn(3);
s(0) <= btn(1) or btn(3);
```

The two OR gates in Fig. 10.1 will implement these logic equations for *s*(1:0).

The constant signal assignment for *x*(15:0) can be made by right clicking on the BDE diagram and selecting *VHDL→signal assignments*. Then, wire the signal assignment box to the *x* input and place the constant assignment *x* <= *X"1234";* in the signal assignment box.

The VHDL program created by compiling *mux7seg.bde* in Fig. 10.1 is equivalent to the VHDL program shown in Listing 10.1. If you implement the design *mux7seg.bde* shown in Fig. 10.1 and download the *.bit* file to the FPGA board, then when you press buttons 0, 1, 2, and 3 the digits 4, 3, 2, and 1 will be displayed on digits 0, 1, 2, and 3 respectively. Try it.

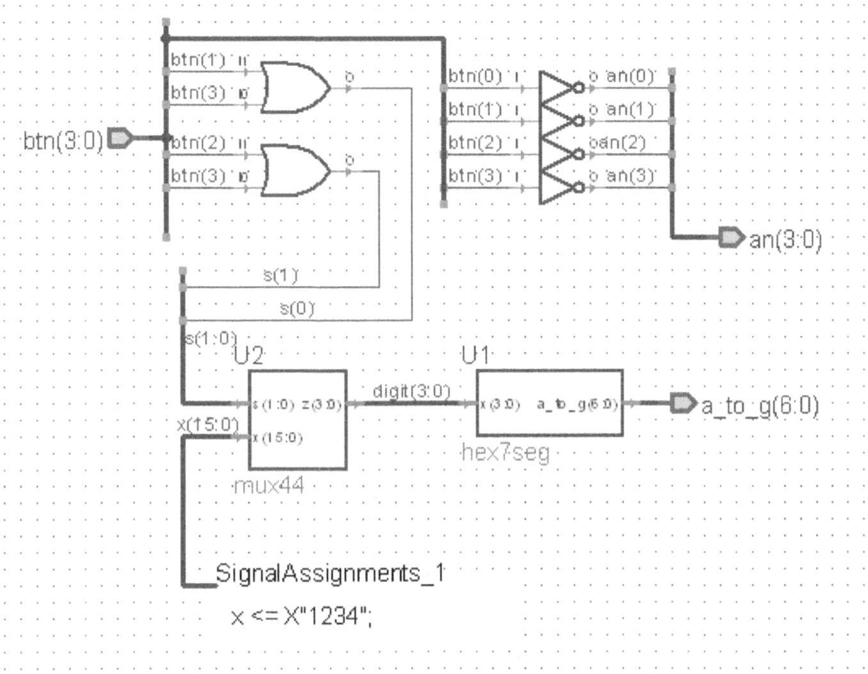

Figure 10.1 BDE circuit *mux7seg.bde* for multiplexing the four 7-segment displays

btn(3)	btn(2)	btn(1)	btn(0)	s(1)	s(0)
0	0	0	0	X	X
0	0	0	1	0	0
0	0	1	0	0	1
0	1	0	0	1	0
1	0	0	0	1	1

Figure 10.2 Truth table for generating s(1:0) in Fig. 10.1

Listing 10.1 mux7seg.vhd

```
-- Example 10a: mux7seg
library IEEE;
use IEEE.std_logic_1164.all;

entity mux7seg is
  port(
        btn : in STD_LOGIC_VECTOR(3 downto 0);
        a_to_g : out STD_LOGIC_VECTOR(6 downto 0);
        an : out STD_LOGIC_VECTOR(3 downto 0)
  );
end mux7seg;
```

Listing 10.1 (cont.) mux7seg.vhd

```
architecture mux7seg of mux7seg is

component hex7seg
  port (
       x : in STD_LOGIC_VECTOR(3 downto 0);
       a_to_g : out STD_LOGIC_VECTOR(6 downto 0)
  );
end component;
component mux44
  port (
       s : in STD_LOGIC_VECTOR(1 downto 0);
       x : in STD_LOGIC_VECTOR(15 downto 0);
       z : out STD_LOGIC_VECTOR(3 downto 0)
  );
end component;

signal digit : STD_LOGIC_VECTOR (3 downto 0);
signal s : STD_LOGIC_VECTOR (1 downto 0);
signal x : STD_LOGIC_VECTOR (15 downto 0);

begin
   x <= X"1234";

   U1 : hex7seg
     port map(
          a_to_g => a_to_g, x => digit);

   U2 : mux44
     port map(
          s => s, x => x, z => digit);

   s(0) <= btn(3) or btn(1);
   s(1) <= btn(3) or btn(2);
   an(1) <= not(btn(1));
   an(0) <= not(btn(0));
   an(2) <= not(btn(2));
   an(3) <= not(btn(3));

end mux7seg;
```

10.2 7-Segment Displays: *x7seg*

We saw in Section 10.1 that to display a 16-bit hex value on the four 7-segment displays we must multiplex the four hex digits. You can only make it appear that all four digits are on by multiplexing them fast enough (greater than 30 times per second) so that your eyes retain the values. This is the same way that your TV works where only a single picture element (pixel) is on at any one time, but the entire screen is refreshed 30 times per second so that you perceive the entire image. To do this the value of $s(1:0)$ in Fig. 10.1 must count from 0 to 3 continually at this fast rate. At the same time the value of the outputs $an(3:0)$ must be synchronized with $s(1:0)$ so as to enable the proper digit at the proper time. A circuit for doing this is shown in Fig. 10.3. The outputs $an(3:0)$ will satisfy the truth table in Fig. 10.4. Note that each output $an(i)$ is just the maxterm $M(i)$ of $q(1:0)$.

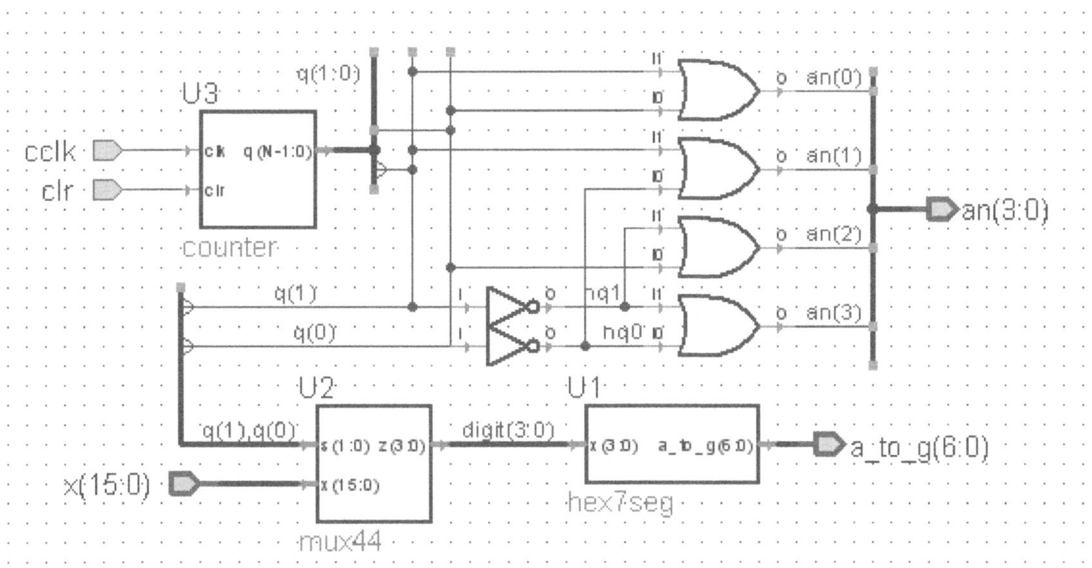

Figure 10.3 BDE circuit *x7seg.bde* for displaying x(15:0) on the four 7-segment displays

q(1) q(0)	an(3) an(2) an(1) an(0)
0 0	1 1 1 0
0 1	1 1 0 1
1 0	1 0 1 1
1 1	0 1 1 1

Figure 10.4 Truth table for generating *an*(3:0) in Fig. 10.3

A simulation of *x7seg.bde* is shown in Fig. 10.5. Note how the *an*(3:0) output selects one digit at a time to display the value 1234 on the 7-segment displays. When *x7seg.bde* is compiled it creates a VHDL program that is equivalent to Listing 10.2. The top-level design shown in Fig. 10.6 can be used to test the *x7seg* module on the FPGA board. The VHDL program corresponding to this top-level design is given in Listing 10.3. Note that the *x7seg* module requires a 190 Hz clock generated by the clock divider module *clkdiv* from Example 8.

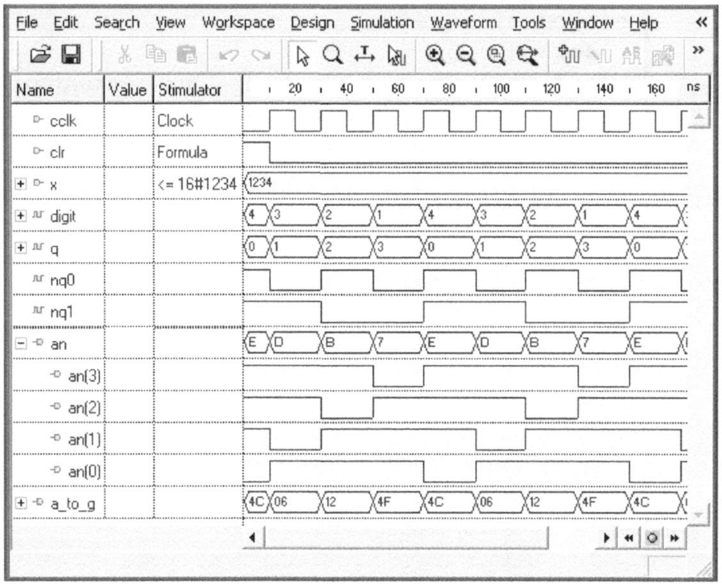

Figure 10.5 Simulation of the *x7segb.bde* circuit in Fig. 10.3

Listing 10.2 x7seg.vhd

```
-- Example 10b: x7seg
library IEEE;
use IEEE.std_logic_1164.all;

entity x7seg is
  port(
       cclk : in STD_LOGIC;
       clr : in STD_LOGIC;
       x : in STD_LOGIC_VECTOR(15 downto 0);
       a_to_g : out STD_LOGIC_VECTOR(6 downto 0);
       an : out STD_LOGIC_VECTOR(3 downto 0)
  );
end x7seg;

architecture x7seg of x7seg is
component counter
  generic(
       N : INTEGER := 8
  );
  port (
       clk : in STD_LOGIC;
       clr : in STD_LOGIC;
       q : out STD_LOGIC_VECTOR(N-1 downto 0)
  );
end component;

component hex7seg
  port (
       x : in STD_LOGIC_VECTOR(3 downto 0);
       a_to_g : out STD_LOGIC_VECTOR(6 downto 0)
  );
end component;

component mux44
  port (
       s : in STD_LOGIC_VECTOR(1 downto 0);
       x : in STD_LOGIC_VECTOR(15 downto 0);
       z : out STD_LOGIC_VECTOR(3 downto 0)
  );
end component;

signal nq0 : STD_LOGIC;
signal nq1 : STD_LOGIC;
signal digit : STD_LOGIC_VECTOR (3 downto 0);
signal q : STD_LOGIC_VECTOR (1 downto 0);

begin
U1 : hex7seg
  port map(
       a_to_g => a_to_g,
       x => digit
  );

nq1 <= not(q(1));
nq0 <= not(q(0));
```

Listing 10.2 (cont.) x7seg.vhd

```
U2 : mux44
  port map(
        s(0) => q(0),
        s(1) => q(1),
        x => x,
        z => digit
  );

U3 : counter
  port map(
        clk => cclk,
        clr => clr,
        q => q( 1 downto 0 )
  );

an(0) <= q(0) or q(1);
an(1) <= nq0 or q(1);
an(2) <= q(0) or nq1;
an(3) <= nq0 or nq1;

end x7seg;
```

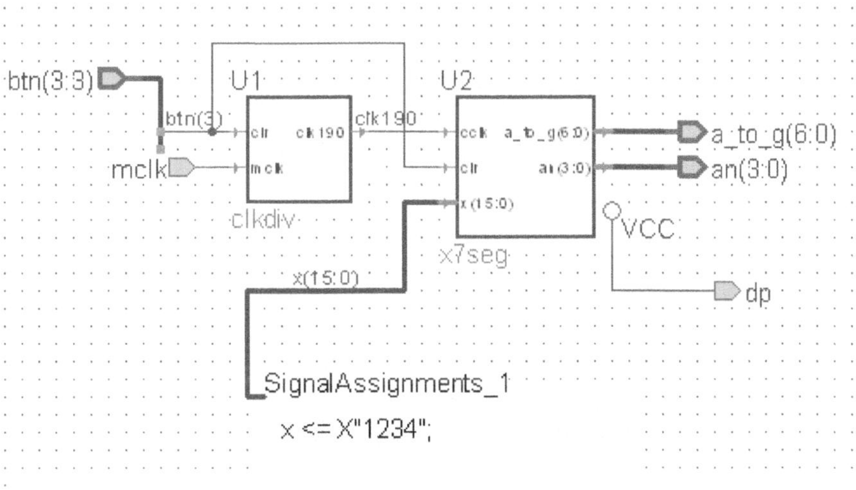

Figure 10.6 Top-level design for testing *x7seg*

Listing 10.3 x7seg_top.vhd

```
-- Example 10c: x7seg_top
library IEEE;
use IEEE.STD_LOGIC_1164.all;

entity x7seg_top is
      port(
             mclk : in STD_LOGIC;
             btn : in STD_LOGIC_VECTOR(3 downto 3);
             a_to_g : out STD_LOGIC_VECTOR(6 downto 0);
             an : out STD_LOGIC_VECTOR(3 downto 0);
             dp : out STD_LOGIC
          );
end x7seg_top;
```

Listing 10.3 (cont.) x7seg_top.vhd

```
architecture x7seg_top of x7seg_top is
component x7seg is
        port(
        cclk : in STD_LOGIC;
        clr : in STD_LOGIC;
        x : in STD_LOGIC_VECTOR(15 downto 0);
        a_to_g : out STD_LOGIC_VECTOR(6 downto 0);
        an : out STD_LOGIC_VECTOR(3 downto 0)
);
end component;
component clkdiv
  port (
        clr : in STD_LOGIC;
        mclk : in STD_LOGIC;
        clk190 : out STD_LOGIC
  );
end component;

signal x: STD_LOGIC_VECTOR(15 downto 0);
signal clk190: STD_LOGIC;

begin

  x <= X"1234";                 -- test display value

  X1: clkdiv port map
     (clr=>btn(3), mclk=>mclk, clk190=>clk190);

  X2: x7seg port map
     (x=>x, cclk=>clk190, clr=>btn(3), a_to_g=>a_to_g,
      an=>an);
  dp <= '1';

end x7seg_top;
```

10.3 7-Segment Displays: *x7segb*

When implementing the circuit for *x7seg* in Fig. 10.3 we must add separate VHDL files to the project for the modules *counter*, *hex7seg* and *mux44*. Alternatively, we can include separate processes within a single VHDL file. A variation of *x7seg*, called *x7segb*, that displays leading zeros as blanks is shown in Listing 10.4. This is done by writing logic equations for *aen*(3:0) that depend on the values of *x*(15:0). For example, *aen*(3) will be 1 (and thus digit 3 will not be blank) if any one of the top four bits of *x*(15:0) is 1. Similarly, *aen*(2) will be 1 if any one of the top eight bits of *x*(15:0) is 1, and *aen*(1) will be 1 if any one of the top twelve bits of *x*(15:0) is 1. Note that *aen*(0) is always 1 so that digit 1 will always be displayed even if it is zero.

To test the module *x7segb* you can run the top-level design shown in Listing 10.4 that will display the value of *x* on the 7-segment displays where *x* is defined by the following statement:

```
x <= sw & btn(2 downto 0) & "01010";   -- digit 0 = A
```

In this case we form the 16-bit value of *x* by concatenating the eight switches, the three right-most pushbuttons, and the five bits 01010. Note that if all switches are off an A will be displayed on digit 0 with three leading blanks. Turning on the switches and pushing the three right-most pushbuttons will display various hex numbers – always with leading blanks.

Listing 10.4 x7segb.vhd

```
-- Example 10d: x7segb - Display 7-seg with leading blanks
-- input cclk should be 190 Hz
library IEEE;
use IEEE.STD_LOGIC_1164.all;
use IEEE.STD_LOGIC_UNSIGNED.all;

entity x7segb is
       port(
              x : in STD_LOGIC_VECTOR(15 downto 0);
              clk : in STD_LOGIC;
              clr : in STD_LOGIC;
              a_to_g : out STD_LOGIC_VECTOR(6 downto 0);
              an : out STD_LOGIC_VECTOR(3 downto 0);
              dp : out STD_LOGIC
           );
end x7segb;

architecture x7segb of x7segb is

signal s: STD_LOGIC_VECTOR(1 downto 0);
signal digit: STD_LOGIC_VECTOR(3 downto 0);
signal aen: STD_LOGIC_VECTOR(3 downto 0);
signal clkdiv: STD_LOGIC_VECTOR(20 downto 0);

begin

   s <= clkdiv(20 downto 19);
   dp <= '1';
   -- set aen(3 downto 0) for leading blanks
   aen(3) <= x(15) or x(14) or x(13) or x(12);
   aen(2) <= x(15) or x(14) or x(13) or x(12)
            or x(11) or x(10) or x(9) or x(8);
   aen(1) <= x(15) or x(14) or x(13) or x(12)
            or x(11) or x(10) or x(9) or x(8)
            or x(7) or x(6) or x(5) or x(4);
   aen(0) <= '1'; -- digit 0 always on

-- Quad 4-to-1 MUX: mux44
   process(s, x)
   begin
      case s is
          when "00" => digit <= x(3 downto 0);
          when "01" => digit <= x(7 downto 4);
          when "10" => digit <= x(11 downto 8);
          when others => digit <= x(15 downto 12);
      end case;
   end process;
```

Listing 10.4 (cont.) x7segb.vhd

```
--7-segment decoder: hex7seg
   process(digit)
   begin
      case digit is
          when X"0" => a_to_g <= "0000001";       --0
          when X"1" => a_to_g <= "1001111";       --1
          when X"2" => a_to_g <= "0010010";       --2
          when X"3" => a_to_g <= "0000110";       --3
          when X"4" => a_to_g <= "1001100";       --4
          when X"5" => a_to_g <= "0100100";       --5
          when X"6" => a_to_g <= "0100000";       --6
          when X"7" => a_to_g <= "0001101";       --7
          when X"8" => a_to_g <= "0000000";       --8
          when X"9" => a_to_g <= "0000100";       --9
          when X"A" => a_to_g <= "0001000";       --A
          when X"B" => a_to_g <= "1100000";       --b
          when X"C" => a_to_g <= "0110001";       --C
          when X"D" => a_to_g <= "1000010";       --d
          when X"E" => a_to_g <= "0110000";       --E
          when others => a_to_g <= "0111000";     --F
      end case;
   end process;

-- Digit select: ancode
   process(s, aen)
   begin
      an <= "1111";
      if aen(conv_integer(s)) = '1' then
         an(conv_integer(s)) <= '0';
      end if;
   end process;

-- Clock divider
   process(clk, clr)
   begin
      if clr = '1' then
         clkdiv <= (others => '0');
      elsif clk'event and clk = '1' then
         clkdiv <= clkdiv + 1;
      end if;
   end process;
end x7segb;
```

Listing 10.5 x7segb_top.vhd

```
-- Example 10e: x7seg_top
library IEEE;
use IEEE.STD_LOGIC_1164.all;

entity x7segb_top is
       port(
              clk : in STD_LOGIC;
              btn : in STD_LOGIC_VECTOR(3 downto 0);
              sw : in STD_LOGIC_VECTOR(7 downto 0);
              a_to_g : out STD_LOGIC_VECTOR(6 downto 0);
              an : out STD_LOGIC_VECTOR(3 downto 0);
              dp : out STD_LOGIC
            );
end x7segb_top;

architecture x7segb_top of x7segb_top is
component x7segb is
       port(
              x : in STD_LOGIC_VECTOR(15 downto 0);
              clk : in STD_LOGIC;
              clr : in STD_LOGIC;
              a_to_g : out STD_LOGIC_VECTOR(6 downto 0);
              an : out STD_LOGIC_VECTOR(3 downto 0);
              dp : out STD_LOGIC
            );
end component;
signal x: STD_LOGIC_VECTOR(15 downto 0);
begin
  -- concatenate switches and 3 buttons
  x <= sw & btn(2 downto 0) & "01010";  -- digit 0 = A

  X2: x7segb port map
      (x=>x,
       clk=>clk,
       clr=>btn(3),
       a_to_g=>a_to_g,
       an=>an,
       dp=>dp
      );

end x7segb_top;
```

Example 11

2's Complement 4-Bit Saturator

In this example we will design a circuit that converts a 6-bit signed number to a 4-bit output that gets saturated at -8 and +7.

Prerequisite knowledge:

Basic Gates – Appendix C
Equality Detector – Example 6
Quad 2-to-1 Multiplexer – Example 6
7-Segment Displays – Example 10

11.1 Creating the Design *sat4bit.bde*

Figure 11.1 shows a circuit called *sat4bit.bde* that was described in the November 2001 issue of NASA Tech Briefs. The circuit will take a 6-bit two's complement number with a signed value between –32 and +31 and convert it to a 4-bit two's complement number with a signed value between –8 and +7. Negative input values less than –8 will be saturated at –8. Positive input values greater than +7 will be saturated at +7.

Note that the two XNOR gates and the AND gate form an equality detector whose output *s* is 1 when *x*(3), *x*(4), and *x*(5) are all equal (see Example 4). This will be the case when the 6-bit input number *x*(5:0) is between -8 and +7. In this case output *y*(3:0) of the quad 2-to-1 MUX will be connected to the input *x*(3:0). If the top three bits of *x*(5:0) are not equal and *x*(5) is 1 then the input value will be less than -8 and the output *y*(3:0) of the quad 2-to-1 MUX will be saturated at -8. On the other hand if the top three bits of *x*(5:0) are not equal and *x*(5) is 0 then the input value will be greater than +7 and the output *y*(3:0) of the quad 2-to-1 MUX will be saturated at +7.

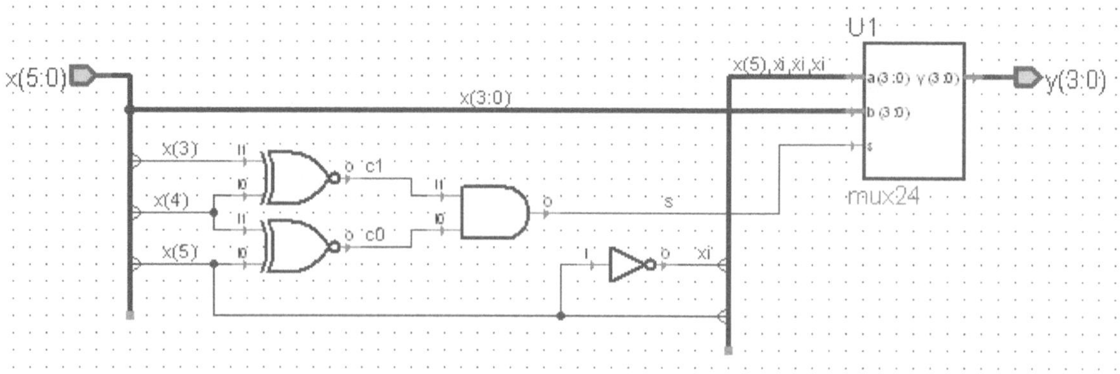

Figure 11.1 Circuit diagram for *sat4bit.bde*

Listing 11.1 sat4bit.vhd

```
-- Example 11a: sat4bit
library IEEE;
use IEEE.std_logic_1164.all;

entity sat4bit is
  port(
        x : in STD_LOGIC_VECTOR(5 downto 0);
        y : out STD_LOGIC_VECTOR(3 downto 0)
  );
end sat4bit;

architecture sat4bit of sat4bit is
component mux24
  port (
        a : in STD_LOGIC_VECTOR(3 downto 0);
        b : in STD_LOGIC_VECTOR(3 downto 0);
        s : in STD_LOGIC;
        y : out STD_LOGIC_VECTOR(3 downto 0)
  );
end component;
signal c0 : STD_LOGIC;
signal c1 : STD_LOGIC;
signal s : STD_LOGIC;
signal xi : STD_LOGIC;

begin
U1 : mux24
  port map(a(0) => xi, a(1) => xi, a(2) => xi, a(3) => x(5),
        b(0) => x(0), b(1) => x(1), b(2) => x(2),
        b(3) => x(3), s => s, y => y);

c1 <= not(x(4) xor x(3));
xi <= not(x(5));
c0 <= not(x(5) xor x(4));
s <= c0 and c1;

end sat4bit;
```

A top-level design that can be used to test *sat4bit* is shown in Fig. 11.2. The module *x7segb11* is a modification of Listing 10.4 that will display only values between -8 and +7 on the 7-segment display. Listing 11.2 shows the VHDL program for the module *x7segb11*. The input to *x7segb11* is the 4-bit output *y*(3:0) from *sat4bit*. Note that only the two rightmost 7-segment display are enabled. The two leftmost displays are always blank. The *hex7seg* process in Listing 11.2 has been modified to display the magnitude of the signed value of y(3:0) – 0 to 8. The preceding 7-segment display will either be blank or display a minus sign. The quad 4-to-1 MUX and the new 2-to-1 MUX are used to display the minus sign when *aen*(1) is enabled if *y*(3) is 1; i.e., if *y* is negative.

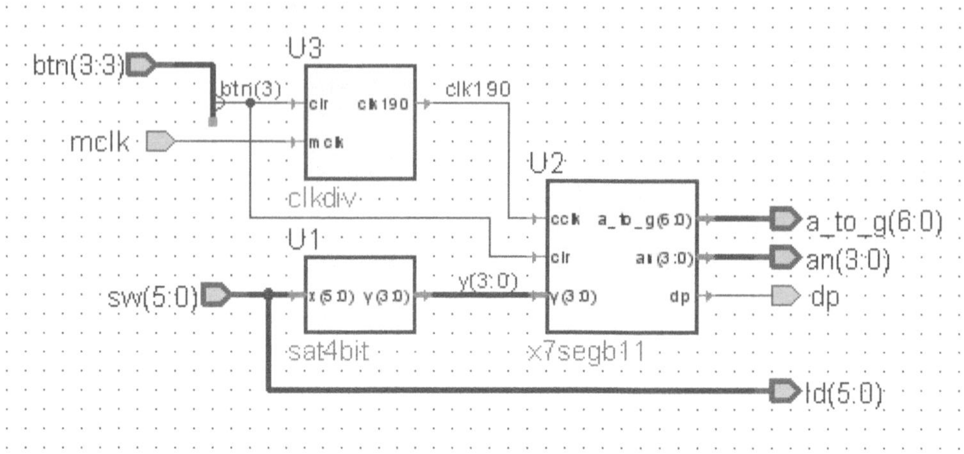

Figure 11.2 Top-level design *sat4bit_top.bde* for testing *sat4bit*

Listing 11.2 x7segb11.vhd

```
-- Example 11b: x7segb11 - test sat4bit
library IEEE;
use IEEE.STD_LOGIC_1164.all;
use IEEE.STD_LOGIC_UNSIGNED.all;

entity x7segb11 is
       port(
              y : in STD_LOGIC_VECTOR(3 downto 0);
              cclk : in STD_LOGIC;
              clr : in STD_LOGIC;
              a_to_g : out STD_LOGIC_VECTOR(6 downto 0);
              an : out STD_LOGIC_VECTOR(3 downto 0);
              dp : out STD_LOGIC
            );
end x7segb11;

architecture x7segb11 of x7segb11 is

signal msel: STD_LOGIC;
signal a_g0: STD_LOGIC_VECTOR(6 downto 0);
signal a_g1: STD_LOGIC_VECTOR(6 downto 0);
signal s: STD_LOGIC_VECTOR(1 downto 0);
signal digit: STD_LOGIC_VECTOR(3 downto 0);
signal aen: STD_LOGIC_VECTOR(3 downto 0);

begin

-- Quad 4-to-1 MUX: mux44
process(s)
begin
  case s is
     when "00" => msel <= '0';
     when "01" => msel <= '1';  --display minus sign
     when "10" => msel <= '0';
     when others => msel <= '0';
  end case;
end process;
```

Listing 11.2 (cont.) x7segb11.vhd

```
--7-segment decoder: hex7seg
process(digit)
begin
  case digit is
     when X"0" => a_g0 <= "0000001";      --0
     when X"1" => a_g0 <= "1001111";      --1
     when X"2" => a_g0 <= "0010010";      --2
     when X"3" => a_g0 <= "0000110";      --3
     when X"4" => a_g0 <= "1001100";      --4
     when X"5" => a_g0 <= "0100100";      --5
     when X"6" => a_g0 <= "0100000";      --6
     when X"7" => a_g0 <= "0001101";      --7
     when X"8" => a_g0 <= "0000000";      -- -8
     when X"9" => a_g0 <= "0000100";      -- -7
     when X"A" => a_g0 <= "0001000";      -- -6
     when X"B" => a_g0 <= "1100000";      -- -5
     when X"C" => a_g0 <= "0110001";      -- -4
     when X"D" => a_g0 <= "1000010";      -- -3
     when X"E" => a_g0 <= "0110000";      -- -2
     when X"F" => a_g0 <= "0110000";      -- -1
     when others => a_g0 <= "0000001";    --0
  end case;
end process;

-- 2-to-1 MUX
process(msel)
begin
      if msel = '1' then
            a_to_g <= a_g1;
      else
            a_to_g <= a_g0;
      end if;
end process;

-- Digit select: ancode
process(s, aen)
begin
      an <= "1111";
      if aen(conv_integer(s)) = '1' then
         an(conv_integer(s)) <= '0';
      end if;
end process;

-- 2-bit counter
process(cclk, clr)
begin
      if clr = '1' then
         s <= "00";
      elsif cclk'event and cclk = '1' then
         s <= s + "01";
      end if;
end process;

end x7segb11;
```

The VHDL program corresponding to the top-level design in Fig. 11.2 is given in Listing 11.3. Download this top-level design to the FPGA board and observe the output on the 7-segment display for different 6-bit switch inputs.

Listing 11.3 sat4bit_top.vhd

```
-- Example 11c: sat4bit_top
library IEEE;
use IEEE.std_logic_1164.all;

entity sat4bit_top is
  port(
        mclk : in STD_LOGIC;
        btn : in STD_LOGIC_VECTOR(3 downto 3);
        sw : in STD_LOGIC_VECTOR(5 downto 0);
        dp : out STD_LOGIC;
        a_to_g : out STD_LOGIC_VECTOR(6 downto 0);
        an : out STD_LOGIC_VECTOR(3 downto 0);
        ld : out STD_LOGIC_VECTOR(5 downto 0)
  );
end sat4bit_top;

architecture sat4bit_top of sat4bit_top is
component clkdiv
  port (
        clr : in STD_LOGIC;
        mclk : in STD_LOGIC;
        clk190 : out STD_LOGIC
  );
end component;

component sat4bit
  port (
        x : in STD_LOGIC_VECTOR(5 downto 0);
        y : out STD_LOGIC_VECTOR(3 downto 0)
  );
end component;

component x7segb11
  port (
        cclk : in STD_LOGIC;
        clr : in STD_LOGIC;
        y : in STD_LOGIC_VECTOR(3 downto 0);
        a_to_g : out STD_LOGIC_VECTOR(6 downto 0);
        an : out STD_LOGIC_VECTOR(3 downto 0);
        dp : out STD_LOGIC
  );
end component;

signal clk190 : STD_LOGIC;
signal y : STD_LOGIC_VECTOR (3 downto 0);
```

Listing 11.3 (cont.) sat4bit_top.vhd

```
begin
U1 : sat4bit
  port map(
        x => sw,
        y => y
  );

U2 : x7segb11
  port map(
        a_to_g => a_to_g,
        an => an,
        cclk => clk190,
        clr => btn(3),
        dp => dp,
        y => y
  );

U3 : clkdiv
  port map(
        clk190 => clk190,
        clr => btn(3),
        mclk => mclk
  );

ld <= sw;

end sat4bit_top;
```

Example 12

Full Adder

In this example we will design a full adder circuit.

Prerequisite knowledge:
Basic Gates – Appendix C
Karnaugh Maps – Appendix D
7-Segment Displays – Example 10

12.1 Half Adder

The truth table for a half adder is shown in Fig. 12.1. In this table bit *a* is added to bit *b* to produce the sum bit *s* and the carry bit *c*. Note that if you add 1 to 1 you get 2, which in binary is 10 or 0 with a carry bit. The BDE logic diagram, *halfadd.bde*, for a half adder is also shown in Fig. 12.1. Note that the sum *s* is just the exclusive-or of *a* and *b* and the carry *c* is just *a* & *b*. The VHDL program corresponding to the circuit in Fig. 12.1 is shown in Listing 12.1. A simulation of *halfadd.bde* is shown in Fig. 12.2.

a	b	s	c
0	0	0	0
0	1	1	0
1	0	1	0
1	1	0	1

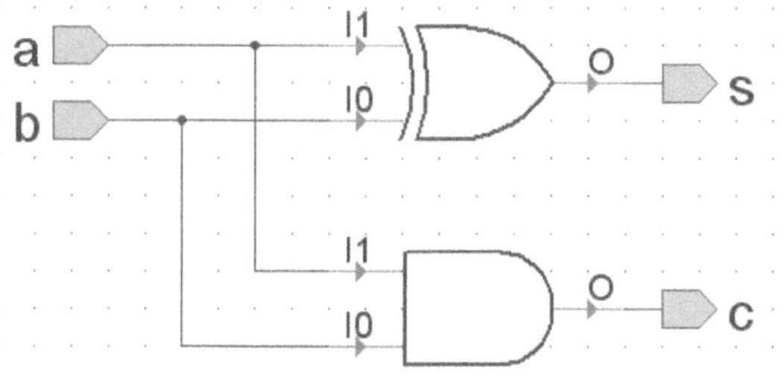

Figure 12.1 Truth table and logic diagram *halfadd.bde* for a half-adder

Listing 12.1 halfadd.vhd

```
-- Example 12a: half adder
library IEEE;
use IEEE.STD_LOGIC_1164.all;
use IEEE.STD_LOGIC_unsigned.all;

entity halfadd is
       port(
             a : in STD_LOGIC;
             b : in STD_LOGIC;
             c : out STD_LOGIC;
             s : out STD_LOGIC
           );
end halfadd;
```

Listing 12.1 (cont.) halfadd.vhd

```
architecture halfadd of halfadd is
begin
        s <= a xor b;
        c <= a and b;

end halfadd;
```

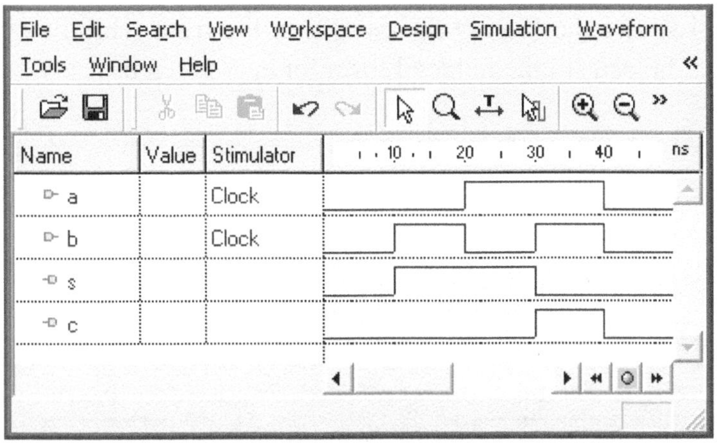

Figure 12.2 Simulation of the half-adder in Fig. 12.1

12.2 Full Adder

When adding binary numbers we need to consider the carry from one bit to the next. Thus, at any bit position we will be adding three bits: a_i, b_i and the carry-in c_i from the addition of the two bits to the right of the current bit position. The sum of these three bits will produce a sum bit, s_i, and a carry-out, c_{i+1}, which will be the carry-in to the next bit position to the left. This is called a *full adder* and its truth table is shown in Fig. 12.3. The results of the first seven rows in this truth table can be inferred from the truth table for the half adder given in Fig. 12.1. In all of these rows only two 1's are ever added together. The last row in Fig. 12.3 adds three 1's. The result is 3, which in binary is 11, or 1 plus a carry.

c_i	a_i	b_i	s_i	c_{i+1}
0	0	0	0	0
0	0	1	1	0
0	1	0	1	0
0	1	1	0	1
1	0	0	1	0
1	0	1	0	1
1	1	0	0	1
1	1	1	1	1

Figure 12.3
Truth table for a full adder

From the truth table in Fig. 12.3 we can write a sum of products expression for s_i as

$$\begin{aligned} s_i = \; & \sim c_i \;\&\; \sim a_i \;\&\; b_i \\ | \; & \sim c_i \;\&\; a_i \;\&\; \sim b_i \\ | \; & c_i \;\&\; \sim a_i \;\&\; \sim b_i \\ | \; & c_i \;\&\; a_i \;\&\; b_i \end{aligned} \qquad (12.1)$$

We can use the distributive law to factor out $\sim c_i$ from the first two product terms and c_i from the last two product terms in Eq. (12.1) to obtain

$$\begin{aligned} s_i = \; & \sim c_i \;\&\; (\sim a_i \;\&\; b_i \;|\; a_i \;\&\; \sim b_i) \\ | \; & c_i \;\&\; (\sim a_i \;\&\; \sim b_i \;|\; a_i \;\&\; b_i) \end{aligned} \qquad (12.2)$$

which can be written in terms of XOR and XNOR operations as

```
si  =  ~ci & (ai ^ bi) | ci & ~(ai ^ bi)                    (12.3)
```

which further reduces to

```
si  =  ci ^ (ai ^ bi)                                        (12.4)
```

Fig. 12.4 shows the K-map for c_{i+1} from the truth table in Fig. 12.3. The map shown in Fig. 12.4a leads to the reduced form for c_{i+1} given by

```
ci+1  =   ai & bi | ci & bi | ci & ai                        (12.5)
```

While this is the reduced form, a more convenient form can be written from Fig. 12.4b as follows:

```
ci+1  = ai & bi | ci & ~ai & bi | ci & ai & ~bi
      = ai & bi | ci & (~ai & bi | ai & ~bi)
      = ai & bi | ci & (ai ^ bi)                             (12.6)
```

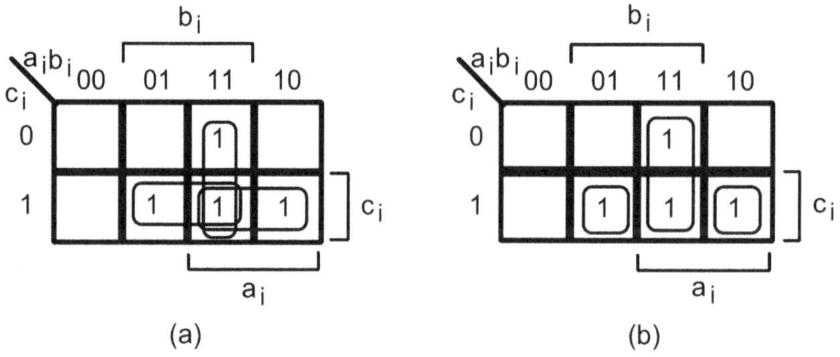

Figure 12.4 K-maps for c_{i+1} for full adder in Fig. 6.2

From Eqs. (12.4) and (12.6) we can draw the logic diagram for a full adder as shown in Fig. 12.5. Comparing this diagram to that for a half adder in Fig. 12.1 it is clear that a full adder can be made from two half adders plus an OR gate as shown in Fig. 12.6.

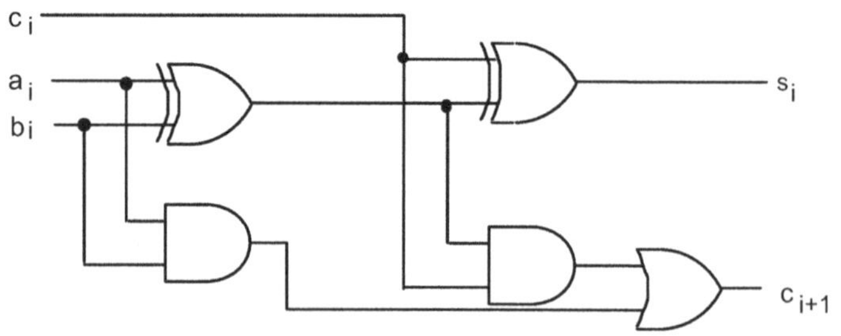

Figure 12.5 Logic diagram for a full adder

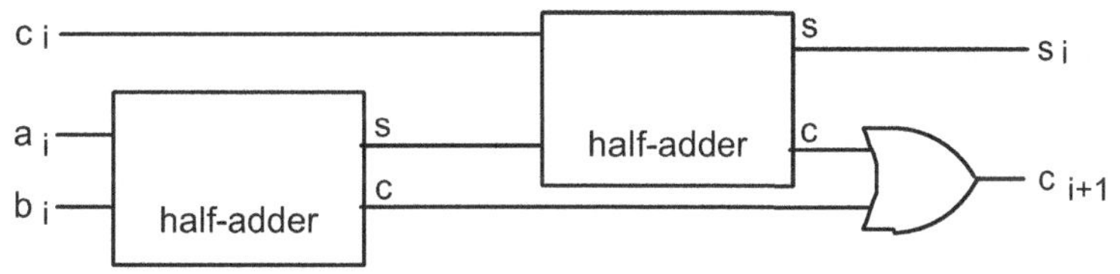

Figure 12.6 A full adder can be made from two half adders plus an OR gate

From Fig. 12.6 we can create a BDE design, *fulladd.bde*, as shown in Fig. 12.7. The VHDL program resulting from compiling this design is equivalent to that shown in Listing 12.2. A simulation of this full adder is shown in Fig. 12.8. Note that the outputs agree with the truth table in Fig. 12.3.

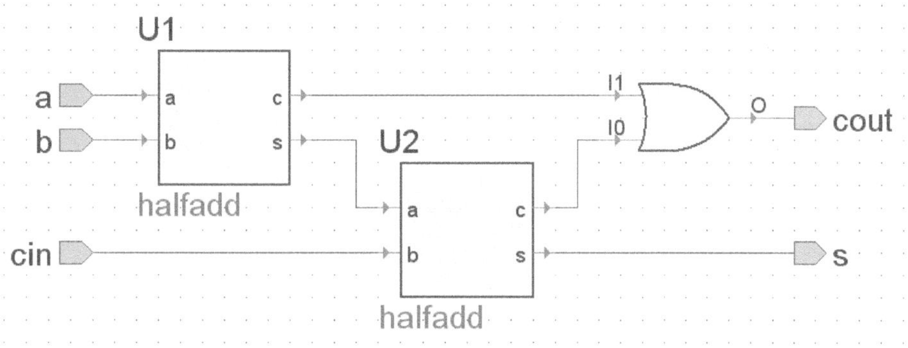

Figure 12.7 Block diagram *fulladd.bde* for a full adder

Listing 12.2 fulladd.v

```
-- Example 12b: fulladd
library IEEE;
use IEEE.std_logic_1164.all;

entity fulladd is
  port(
        a : in STD_LOGIC;
        b : in STD_LOGIC;
        cin : in STD_LOGIC;
        cout : out STD_LOGIC;
        s : out STD_LOGIC
  );
end fulladd;

architecture fulladd of fulladd is
component halfadd
  port (
        a : in STD_LOGIC;
        b : in STD_LOGIC;
        c : out STD_LOGIC;
        s : out STD_LOGIC
  );
end component;
```

Listing 12.2 (cont.) fulladd.v

```
signal c1 : STD_LOGIC;
signal c2 : STD_LOGIC;
signal s1 : STD_LOGIC;

begin
U1 : halfadd
  port map(
        a => a, b => b, c => c1, s => s1);

U2 : halfadd
  port map(
        a => s1, b => cin, c => c2, s => s);

cout <= c2 or c1;

end fulladd;
```

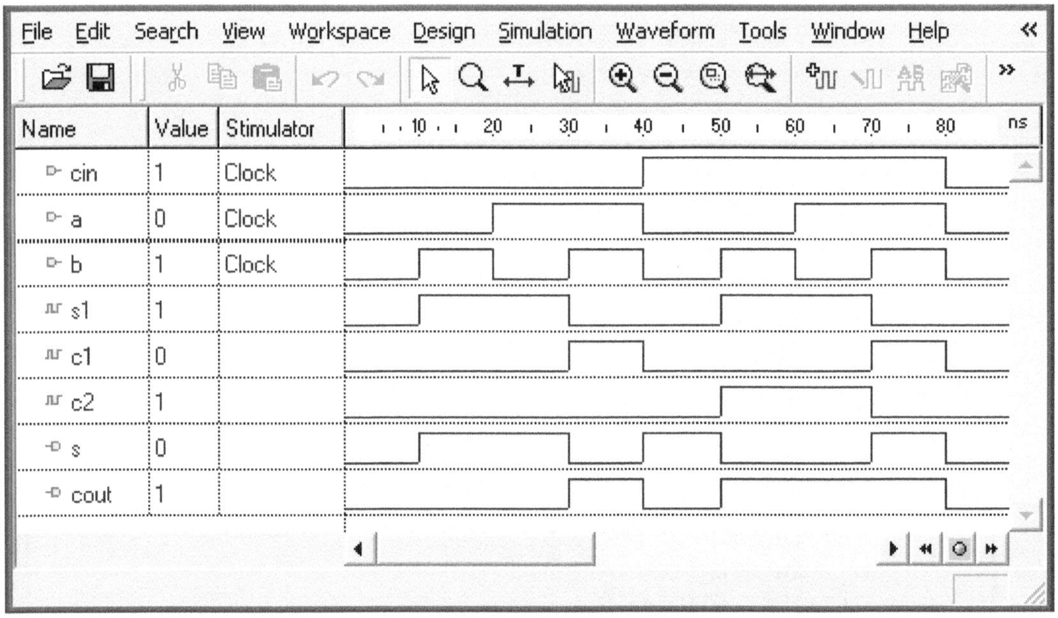

Figure 12.8 Simulation of the full adder in Fig. 12.7 and Listing 12.2

Example 13

4-Bit Adder

In this example we will design a 4-bit adder.

Prerequisite knowledge:
Basic Gates – Appendix C
Karnaugh Maps – Appendix D
Full Adder – Example 12

13.1 4-Bit Adder

Four of the full adders in Fig. 12.7 can be combined to form a 4-bit adder as shown in Fig. 13.1. Note that the full adder for the least significant bit will have a carry-in of zero while the remaining bits get their carry-in from the carry-out of the previous bit. The final carry-out, is the *cout* for the 4-bit addition. The VHDL program corresponding to the 4-bit adder in Fig. 13.1 is given in Listing 13.1.

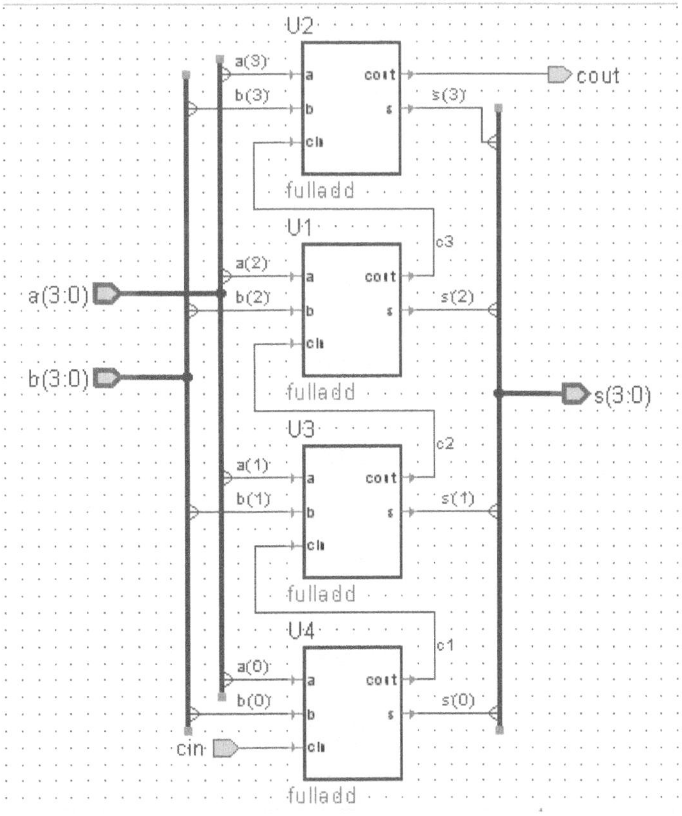

Figure 13.1 Block diagram *adder4.bde* for a 4-bit adder

Listing 13.1 adder4.vhd

```
-- Example 13a: adder4
library IEEE;
use IEEE.std_logic_1164.all;

entity adder4 is
  port(
       cin : in STD_LOGIC;
       a : in STD_LOGIC_VECTOR(3 downto 0);
       b : in STD_LOGIC_VECTOR(3 downto 0);
       cout : out STD_LOGIC;
       s : out STD_LOGIC_VECTOR(3 downto 0)
  );
end adder4;

architecture adder4 of adder4 is
component fulladd
  port (
       a : in STD_LOGIC;
       b : in STD_LOGIC;
       cin : in STD_LOGIC;
       cout : out STD_LOGIC;
       s : out STD_LOGIC
  );
end component;

signal c1 : STD_LOGIC;
signal c2 : STD_LOGIC;
signal c3 : STD_LOGIC;

begin
U1 : fulladd
  port map(
       a => a(2), b => b(2), cin => c2, cout => c3,
       s => s(2));

U2 : fulladd
  port map(
       a => a(3), b => b(3), cin => c3, cout => cout,
       s => s(3));

U3 : fulladd
  port map(
       a => a(1), b => b(1), cin => c1, cout => c2,
       s => s(1));

U4 : fulladd
  port map(
       a => a(0), b => b(0), cin => cin, cout => c1,
       s => s(0));

end adder4;
```

A simulation of the 4-bit adder in Fig. 13.1 and Listing 13.1 is shown in Fig. 13.2. The value of *a* is incremented from 0 to F and is added to the hex value B. The sum *s* is always equal to *a* + *b*. Note that the carry flag, *cout*, is equal to 1 when the correct *unsigned* answer exceeds 15 (or F).

We can test the *adder4* module from Fig. 13.1 and Listing 13.1 on the FPGA board by combining it with the *x7segb* module from Listing 10.4 in Example 10 and the *clkdiv* module from Listing 8.2 from Example 8 to produce the top-level design shown in Listing 13.2. The 4-bit number *sw*(7:4) will be displayed on the first (left-most) 7-segment display. The 4-bit number *sw*(3:0) will be displayed on the second 7-segment display. These two numbers will be added and the 4-bit sum will be displayed on the fourth (right-most) 7-segment display and the carry bit will be displayed on the third 7-segment display. Try it.

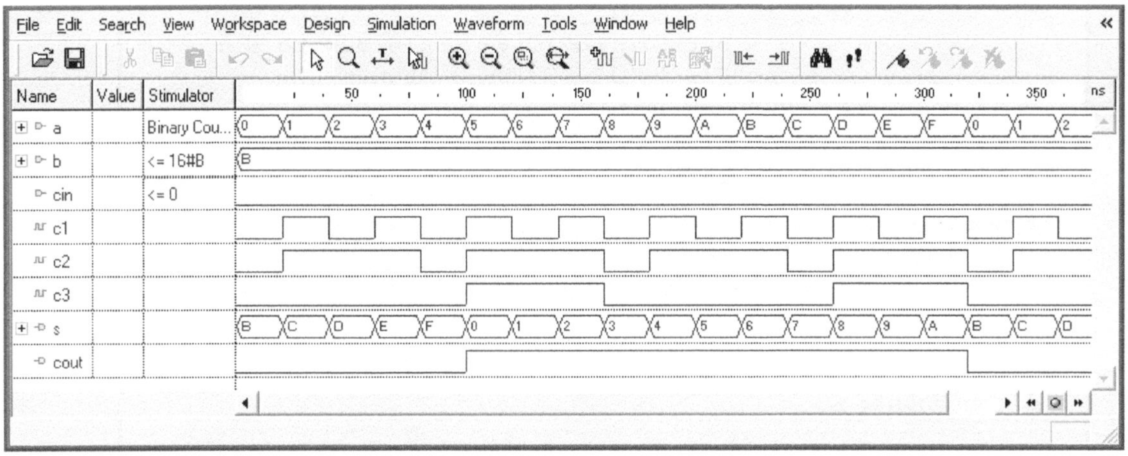

Figure 13.2 Simulation of the 4-bit adder in Fig. 13.1 and Listing 13.1

Listing 13.2 adder4_top.vhd

```
-- Example 13b: adder4_top
library IEEE;
use IEEE.STD_LOGIC_1164.all;

entity adder4_top is
       port(
              mclk : in STD_LOGIC;
              btn : in STD_LOGIC_VECTOR(3 downto 3);
              sw : in STD_LOGIC_VECTOR(7 downto 0);
              a_to_g : out STD_LOGIC_VECTOR(6 downto 0);
              an : out STD_LOGIC_VECTOR(3 downto 0);
              dp : out STD_LOGIC;
              ld : out STD_LOGIC_VECTOR(7 downto 0)
           );
end adder4_top;
```

Listing 13.2 (cont.) adder4_top.vhd

```
architecture adder4_top of adder4_top is
component adder4 is
  port(
       cin : in STD_LOGIC;
       a : in STD_LOGIC_VECTOR(3 downto 0);
       b : in STD_LOGIC_VECTOR(3 downto 0);
       cout : out STD_LOGIC;
       s : out STD_LOGIC_VECTOR(3 downto 0)
  );
end component;

component x7segb is
       port(
             x : in STD_LOGIC_VECTOR(15 downto 0);
             clk : in STD_LOGIC;
             clr : in STD_LOGIC;
             a_to_g : out STD_LOGIC_VECTOR(6 downto 0);
             an : out STD_LOGIC_VECTOR(3 downto 0);
             dp : out STD_LOGIC
           );
end component;

component clkdiv2 is
       port(
             mclk : in STD_LOGIC;
             clr : in STD_LOGIC;
             clk190 : out STD_LOGIC
           );
end component;

signal clk190, clr, c4, cin: STD_LOGIC;
signal x: STD_LOGIC_VECTOR(15 downto 0);
signal sum: STD_LOGIC_VECTOR(3 downto 0);

begin
      cin <= '0';
      x <= sw & "000" & c4 & sum;
      clr <= btn(3);
      ld <= sw;

U1: adder4 port map
    (cin => cin, a => sw(7 downto 4), b => sw(3 downto 0),
     cout => c4, s => sum);

U2: x7segb port map
    (x => x, clk => clk190, clr => clr, a_to_g => a_to_g,
     an => an, dp => dp);

U3: clkdiv2 port map
    (mclk => mclk, clr => clr, clk190 => clk190);

end adder4_top;
```

Example 14

N-Bit Adder

In this example we will design a *N*-bit adder.

Prerequisite knowledge:
4-Bit Adder – Example 13

14.1 4-Bit Adder: Behavioral Statements

It would be convenient to be able to make a 4-bit adder (or any size adder) by just using a + sign in a VHDL statement. In fact, we can! When you write $a + b$ in a VHDL program the compiler will produce a full adder of the type we designed in Example 12. The only question is how to create the output carry bit. The trick is to add a leading 0 to *a* and *b* and then make a 5-bit temporary variable to hold the sum as shown in Listing 14.1. The most-significant bit of this 5-bit sum will be the *carry flag*.

A simulation of this program is shown in Fig. 14.1. Compare this with Fig. 13.2.

Listing 14.1 adder4b.vhd

```
-- Example 14a:  4-bit behavioral adder
library IEEE;
use IEEE.STD_LOGIC_1164.all;
use IEEE.STD_LOGIC_unsigned.all;

entity adder4b is
       port(
             a : in STD_LOGIC_VECTOR(3 downto 0);
             b : in STD_LOGIC_VECTOR(3 downto 0);
             s : out STD_LOGIC_VECTOR(3 downto 0);
             cf : out STD_LOGIC
           );
end adder4b;

architecture adder4b of adder4b is
begin
   process(a, b)
      variable temp: STD_LOGIC_VECTOR(4 downto 0);
   begin
      temp := ('0' & a) + ('0' & b);
      s <= temp(3 downto 0);
      cf <= temp(4);
      s <= a + b;
   end process;

end adder4b;
```

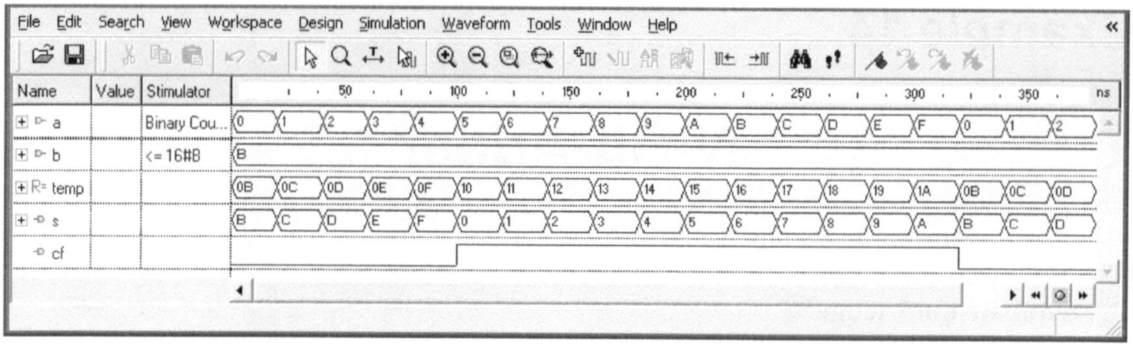

Figure 14.1 Simulation of the VHDL program in Listing 14.1

14.2 *N* - Bit Adder: Behavioral Statements

Listing 14.2 shows an *N*-bit adder that uses a *generic* statement. This is a convenient adder to use when you don't need the carry flag. An example of using this as an 8-bit adder is shown in the simulation in Fig. 14.2. Note that when the sum exceeds FF it simply wraps around and the carry flag is lost.

Listing 14.2 adder.vhd

```
-- Example 14b:  N-bit adder
library IEEE;
use IEEE.STD_LOGIC_1164.all;
use IEEE.STD_LOGIC_unsigned.all;

entity adder is
       generic (N:integer := 8);
       port(
              a : in STD_LOGIC_VECTOR(N-1 downto 0);
              b : in STD_LOGIC_VECTOR(N-1 downto 0);
              y : out STD_LOGIC_VECTOR(N-1 downto 0)
);
end adder;

architecture adder of adder is
begin
   process(a, b)
   begin
      y <= a + b;
   end process;

end adder;
```

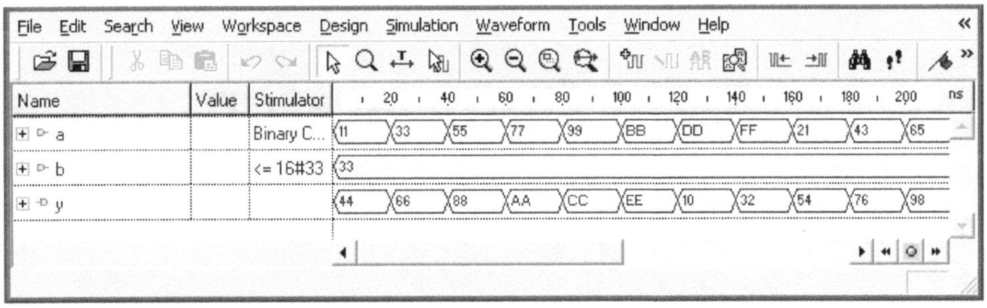

Figure 14.2 Simulation of the VHDL program in Listing 14.2

The top-level design shown in Fig. 14.3 can be used to test this *N*-bit adder on the FPGA board. In this case we are adding two 4-bit switch settings and observing the sum on the 7-segment display. To set the parameter *N* to 4 right-click on the adder symbol, select *Properties* and click on the *Parameter* tab. Set the actual value of *N* to 4.

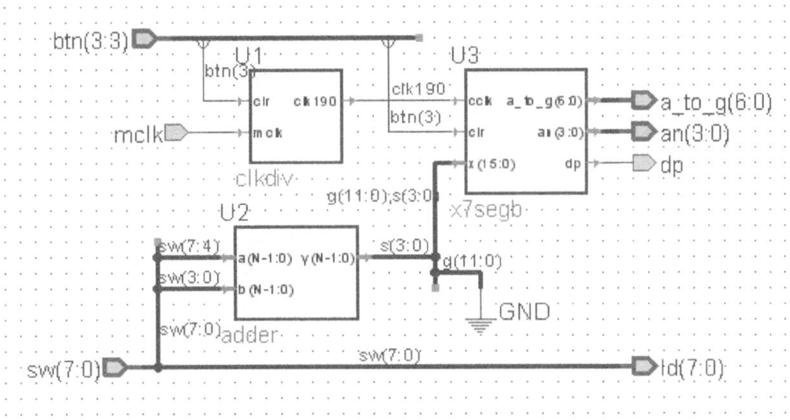

Figure 14.3 Top-level design for testing the *N*-bit adder on the FPGA board

Example 15

N-Bit Comparator

In this example we will design a *N*-bit comparator.

Prerequisite knowledge:
N-Bit Adder – Example 14

15.1 *N*-Bit Comparator Using Relational Operators

The easiest way to implement a comparator in VHDL is to use the relational and logical operators shown in Table 15.1. An example of using these to implement an *N*-bit comparator is shown in Listing 15.1. A simulation of this program for the default value of $N = 8$ is shown in Fig. 15.1.

Note in the *process* in Listing 15.1 we set the values of *gt*, *eq*, and *lt* to zero before the *if* statements. This is important to make sure that each output has a value assigned to it. If you don't do this then VHDL will assume you don't want the value to change and will include a latch in your system. Your circuit will then not be a combinational circuit.

Table 15.1 Relational and Logical Operators

Operator	Meaning
=	Logical equality
/=	Logical inequality
<	Less than
<=	Less than or equal
>	Greater than
>=	Greater than or equal
`not`	Logical negation
`and`	Logical AND
`or`	Logical OR

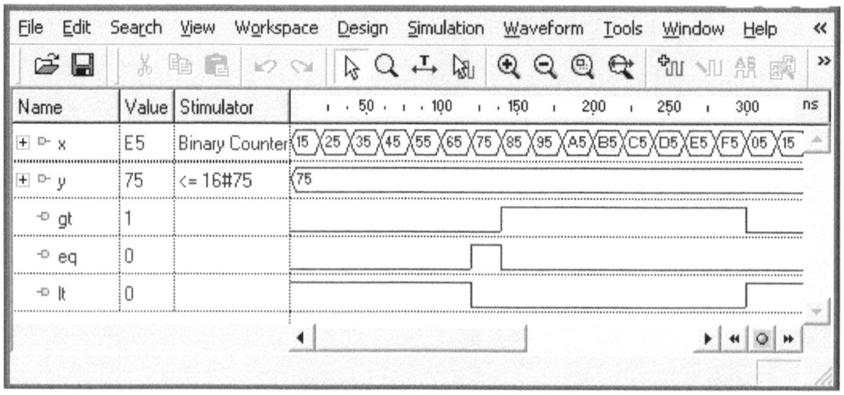

Figure 15.1 Simulation of the VHDL program in Listing 15.1

Listing 15.1 comp.vhd

```
-- Example 17: N-bit comparator using relational operators
library IEEE;
use IEEE.STD_LOGIC_1164.all;

entity comp is
    generic (N:integer := 8);
    port(
           x : in STD_LOGIC_VECTOR(N-1 downto 0);
           y : in STD_LOGIC_VECTOR(N-1 downto 0);
           gt : out STD_LOGIC;
           eq : out STD_LOGIC;
           lt : out STD_LOGIC
        );
end comp;

architecture comp of comp is
begin
   process(x, y)
   begin
         gt <= '0';
         eq <= '0';
         lt <= '0';
         if (x > y) then
                gt <= '1';
         elsif (x = y) then
                eq <= '1';
         elsif (x < y) then
                lt <= '1';
         end if;
   end process;
end comp;
```

You can test this comparator on the FPGA board by creating the BDE block diagram *comp4_top.bde* shown in Fig. 15.2. To make this a 4-bit comparator right-click on the *comp* symbol, select *Properties*, click on the *Parameters* tab, and set the actual value of *N* to 4. You will be comparing the 4-bit number *x*(3:0) on the left four switches with the 4-bit number *y*(3:0) on the right four switches. The three LEDs *ld*(4:2) will detect the outputs *gt*, *eq*, and *lt*. We selected these three LEDs because on the BASYS board they are three different colors. Compile the design *comp4_top.bde*, implement it, and download the *.bit* file to the FPGA board. Test the comparator by changing the switch settings.

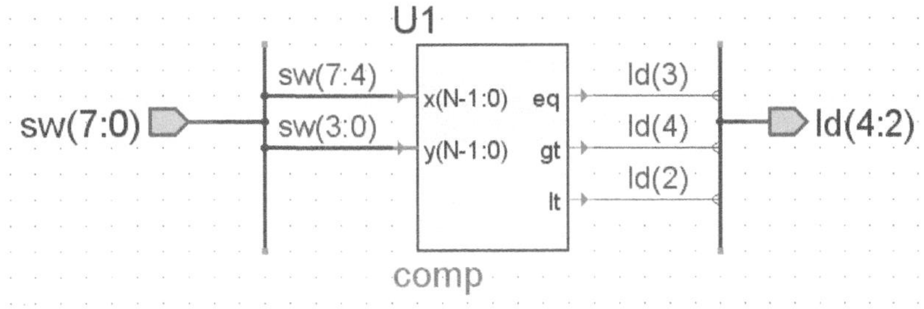

Figure 15.2 Top-level design *comp4_top.bde* to test a 4-bit comparator

Example 16

Edge-Triggered D Flip-Flop

In this example we will define an edge-triggered D flip-flop and show how to design one using only NAND gates.

Prerequisite knowledge:
Basic Gates – Appendix C
Example 8 – Clocks and Counters

16.1 Edge-Triggered D Flip-Flop with Set and Clear

The logic diagram and truth table for a positive edge-triggered D flip-flop with asynchronous set and reset inputs are shown in Fig. 16.1. The upward arrow for the *Clk* signal in the truth table indicates that it is a positive edge-triggered flip-flop. This is also indicated by the arrow-type symbol next to the *CLK* input in the logic diagram. A negative edge-triggered flip-flop would have a bubble added to the *CLK* input.

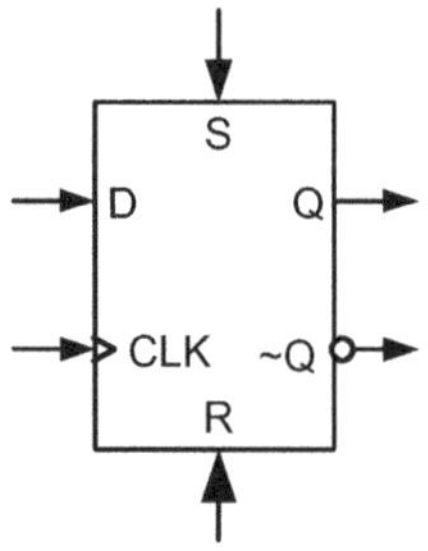

S	R	D	Clk	Q	~Q
0	0	0	↑	0	1
0	0	1	↑	1	0
1	0	X	X	1	0
0	1	X	X	0	1
0	0	X	0	Q_0	$\sim Q_0$

Figure 16.1 Logic diagram for an edge-triggered D flip-flop with asynchronous set and reset

Note that the behavior of a *positive edge-triggered D flip-flop* is that the value of *D* gets latched to *Q* on the rising edge of the clock, *CLK*. If the set input *S* is 1 the output *Q* is set to 1 asynchronously, i.e., regardless of the value of the clock input. If the reset input *R* is 1 the output *Q* is cleared to 0 asynchronously. If both *S* and *R* are zero then the output *Q* changes only on the rising edge of the clock – and is set to the current value of the input D.

There are many ways to make a D flip-flop. The circuit shown in Fig. 16.2 that contains six NAND gates will behave like a D flip-flop with asynchronous set and clear. You can create this D flip-flop using the Active-HDL BDE. Figure 16.3 shows the simulation of this *dff.bde* design. Note from this simulation that the circuit does indeed behave like a D flip-flop in which the output *q* gets latched to the value of *D* on the rising edge of the clock, *clk*. By following all of the internal signals in this simulation you can trace exactly how this circuit works.

We saw in Fig. 1 of the *Introduction* that each of the four slices in each CLB of a Spartan-3E FPGA contains two D flip-flops. The BASYS board contains a Xilinx Spartan3E-100 TQ144 FPGA with 960 slices and 1,920 flip-flops. The Nexys-2 board contains a Xilinx Spartan3E-500 FG320 FPGA with 4,656 slices and 9,312 flip-flops. Each I/O block connected to the FPGA pins contains two additional flip-flops. Thus you have thousands of flip-flops available for you to use in your FPGA designs.

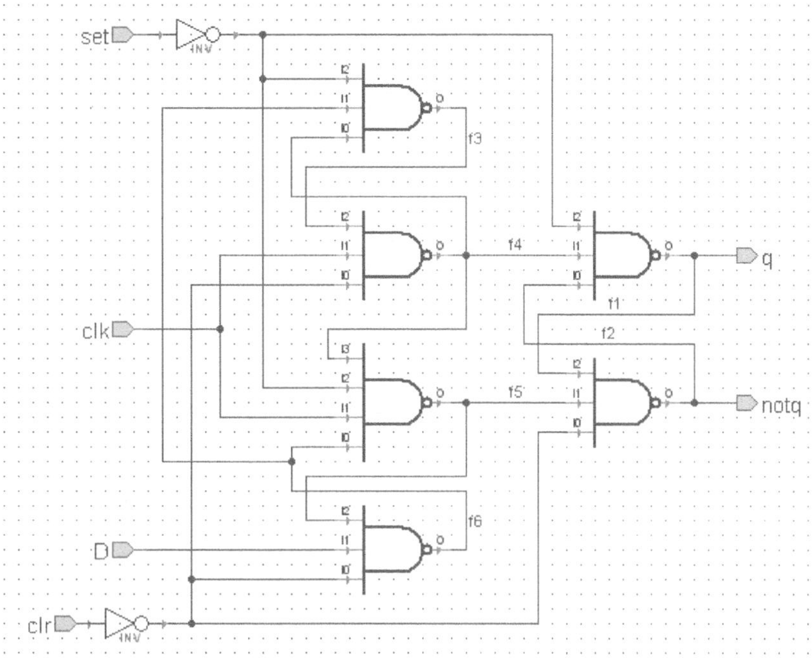

Figure 16.2 Making an edge-triggered D flip-flop with asynchronous set and reset using NAND gates

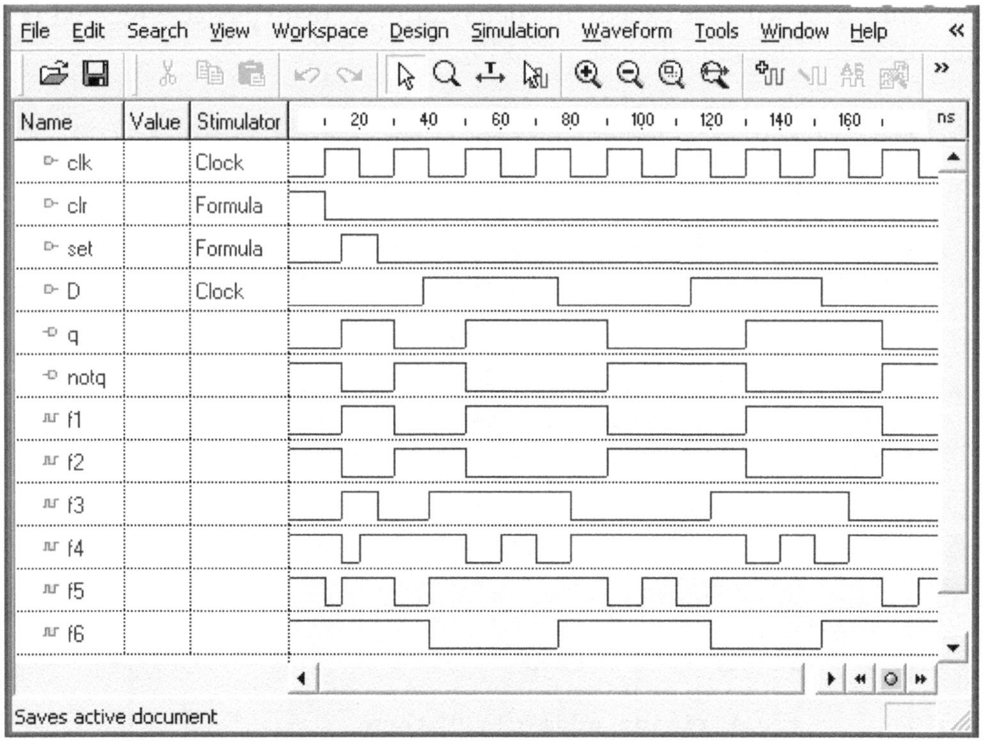

Figure 16.3 Simulation of the D flip-flop *dff.bde* in Fig. 16.2

Example 17

D Flip-Flops in VHDL

In this example we will show how to implement a D flip-flop using VHDL.

Prerequisite knowledge:
Example 8 – Clocks and Counters
Example 16 – Edge-Triggered D Flip-Flop

17.1 D Flip-Flops in VHDL

In Example 16 we showed that an edge-triggered D flip-flop could be implemented using only NAND gates and that FPGAs contain thousands of D flip-flops that are implemented in a variety of specialized ways. These D flip-flops normally have an asynchronous set and clear and behave the same as the simulation in Fig. 16.3 in Example 16.

The way to tell VHDL that you want a positive edge-triggered D flip-flop is to describe its behavior in a process that includes an *if* statement containing the phrase *rising_edge*(*clk*) or *clk'event* ***and*** *clock* = '1' in the *elsif* clause as shown in Listing 17.1. The use of either of these phrases is what tells VHDL to use the D flip-flop in the FPGA. Note that if *clr* is equal to '1' then *q* will immediately (asynchronously) become '0'. If *clr* is not '1' then on the rising edge of the clock signal, *clk*, the value of *q* will be set to the current value of *D*, which is exactly the behavior of a positive edge-triggered D flip-flop. A simulation of Listing 17.1 is shown in Fig. 17.1.

Listing 17.1 Dff.vhd

```
-- Example 17: D flip-flop with clear
library IEEE;
use IEEE.STD_LOGIC_1164.all;

entity Dff is
       port(
              clk : in STD_LOGIC;
              clr : in STD_LOGIC;
              D : in STD_LOGIC;
              q : out STD_LOGIC
            );
end Dff;

architecture Dff of Dff is
begin
  process(clk, clr)
  begin
      if(clr = '1') then
            q <= '0';
      elsif(rising_edge(clk)) then
            q <= D;
      end if;
  end process;
end Dff;
```

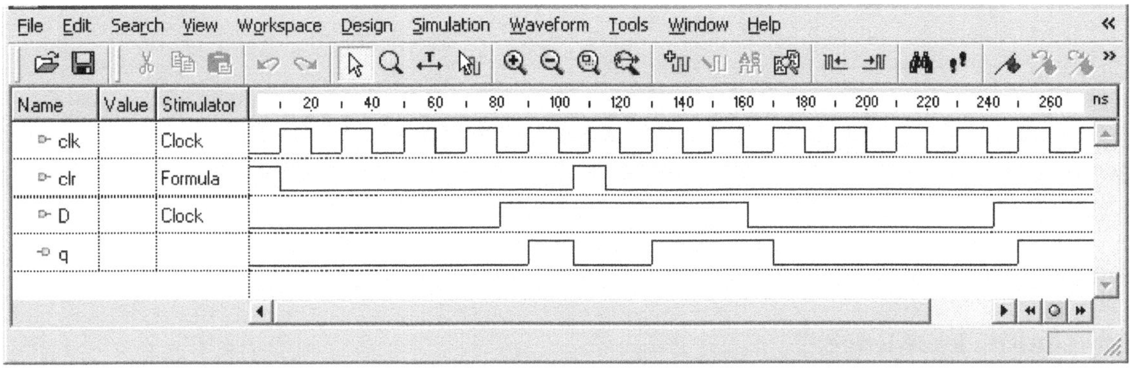

Figure 17.1 Simulation of the VHDL program in Listing 17.1

The BDE symbol for the D flip-flop described in Listing 17.1 is shown in Fig. 17.2. We will show in Example 18 how to use this D flip-flop to create a divide-by-2 counter.

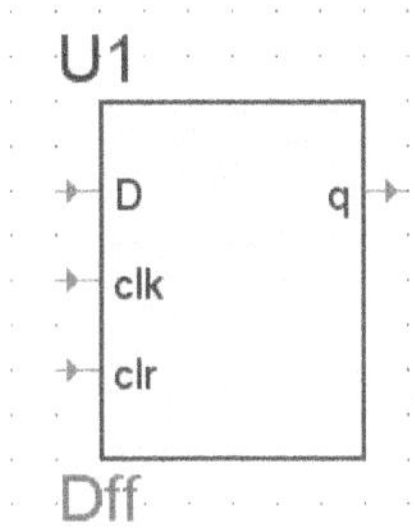

Figure 17.2 BDE symbol for the D flip-flop described in Listing 17.1

Example 18

Divide-By-2 Counter

In this example we will show how to use a D flip-flop to create a divide-by-2 counter.

Prerequisite knowledge:
Example 8 – Clocks and Counters
Example 17 –D Flip-Flops in VHDL

18.1 Divide-by-2 Counter

Consider the D flip-flop with output *q*0 shown in Fig. 18.1 where we have connected ~*q*0 to the *D* input of the flip-flop. What will happen? On each rising edge of the clock, the value at *D* will be latched (after some propagation delay) to *q*0. Suppose that *q*0 is initially 0. This means that ~*q*0 will be 1. At the first clock, this ~*q0* value of 1 (which is at *D*) will be latched to *q*0. Therefore, *q*0 will go from 0 to 1 and ~*q*0 will go from 1 to 0. This means that *D* will now be 0 so that on the next rising edge of the clock, the value of *q*0 will go back to 0, and ~*q*0 will go back to 1. This process will continually repeat itself, resulting in the frequency of *q*0 being just one-half of the frequency of the clock. We call this a *divide-by-2 counter* because it divides the frequency of the clock by two.

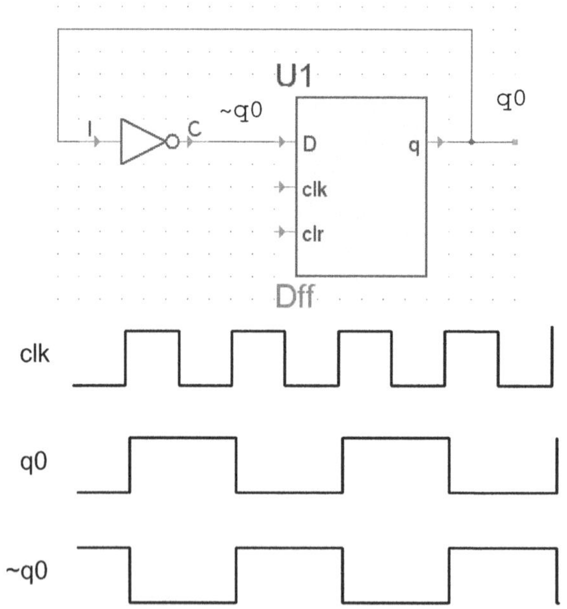

Figure 18.1 Using a D flip-flop as a divide-by-2 counter

A top-level BDE design for this divide-by-2 counter is shown in Fig. 18.2 where we have used the clock divider from Example 8. We will use a 3 Hz clock output *clk*3 from the clock divider. This clock signal will go to *ld*(3) and the ouput of the divide-by-2 counter will go to *ld*(2). A simulation of this divide-by-2 counter is shown in Fig. 18.3. If you implement this design on the BASYS FPGA board the green LED, *ld*(2), will blink at half the frequency of the yellow LED, *ld*(3). Try it.

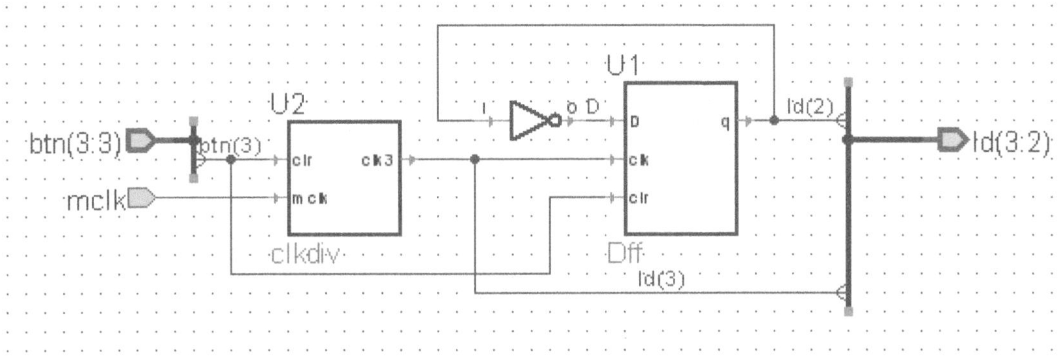

Figure 18.2 Top-level design of the divide-by-2 counter

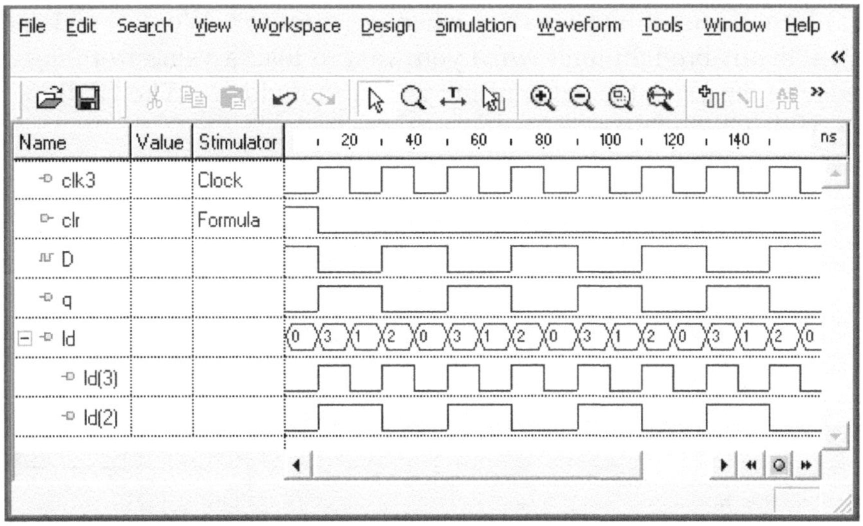

Figure 18.3 Simulation of the divide-by-2 counter in Fig. 18.2

Example 19

Registers

In this example we will show how to use a D flip-flop to create a 1-bit register. We will then use two of these 1-bit registers to form a 2-bit register.

Prerequisite knowledge:
Example 8 – Clocks and Counters
Example 17 –D Flip-Flops in VHDL

19.1 1-Bit Register

In Example 17 we saw that the D flip-flop can be used to store a bit. If *D* is high then on the rising edge of the clock the output *q* of the D flip-flop will become 1. If *D* is low then on the rising edge of the clock the output *q* of the D flip-flop will become 0. In real digital systems the clock input to a D flip-flop is normally on all the time. This means that on every rising edge of the clock (usually millions of times per second) the current value of *D* will be latched to *q*. How can we make a 1-bit register that will load a value (0 or 1) from an input line *inp* only when we want to? We will add another input line called *load* that is brought high when you want to load a value from *inp*, and on the next rising edge of the clock the value of *inp* will be stored in *q*. The BDE logic diagram shown in Fig. 19.1 will do this.

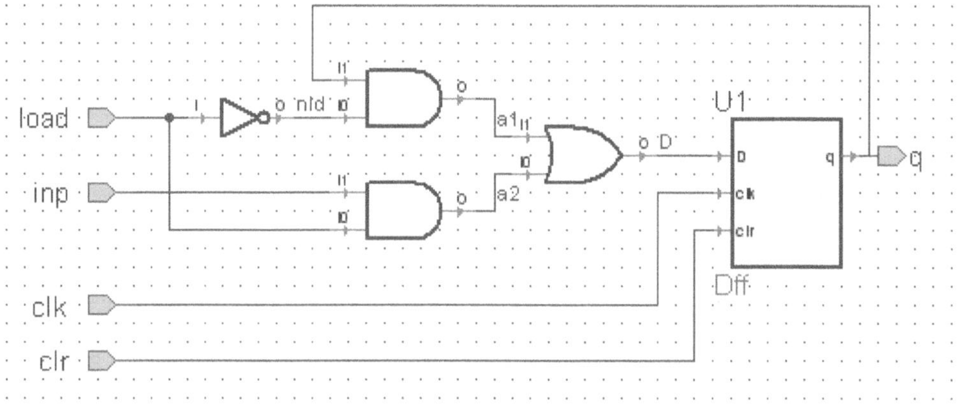

Figure 19.1 A 1-bit register *reg1bit.bde*

The clock signal is assumed to be running continuously and so in order to keep the current value of *q* unchanged at each clock cycle, this *q* value is fed back and gated into the OR gate with *~load.* This means that when the *load* signal is *LO* the value of *q* is continually reloaded and therefore does not change. When *load* is brought *HI* the value *inp* is gated to the OR gate so that on the next clock cycle *q* becomes equal to *inp.* When the *load* signal goes *LO* again this value then remains at *q*. A simulation of *reg1bit.bde* is shown in Fig. 19.2.

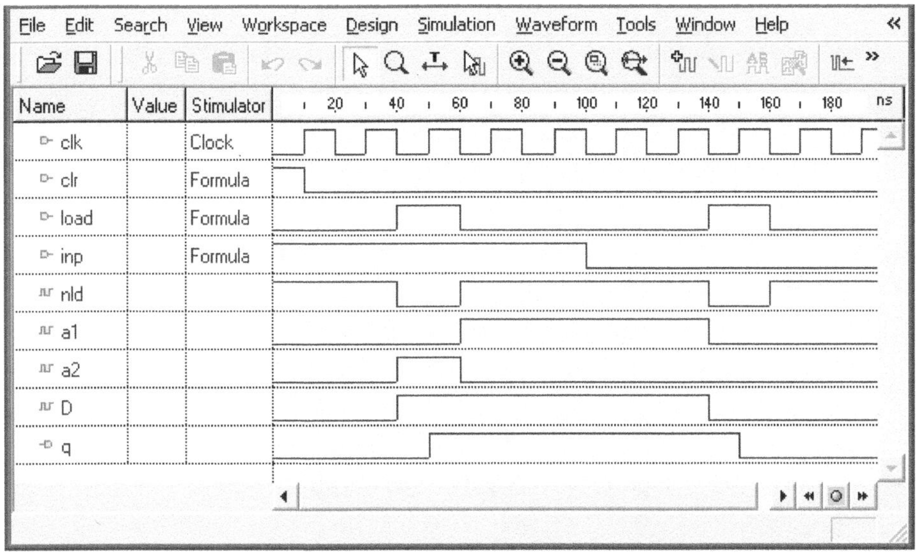

Figure 19.2 Simulation of the 1-bit register *reg1bit.bde* in Fig. 19.1

The logic symbol for this 1-bit register is shown in Fig. 19.3. In the next section we will combine two of these 1-bit register modules to form a 2-bit register.

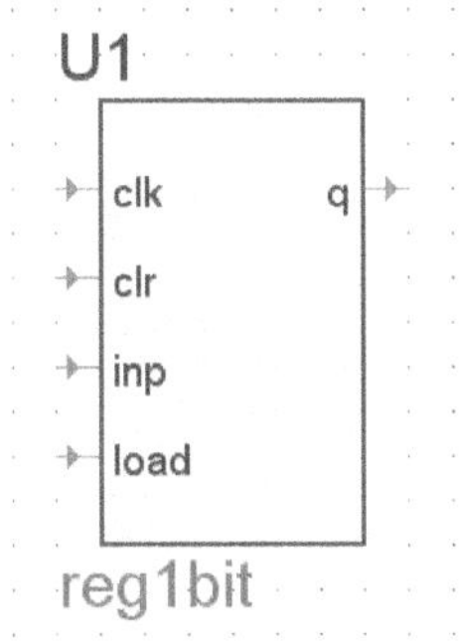

Figure 19.3 Logic symbol for a 1-bit register

19.2 2-Bit Register

We can combine two of the 1-bit register modules shown in Fig. 19.3 with common *load*, *clr*, and *clk* signals to implement a 2-bit register called *reg2bit* as shown in Fig. 19.4. A simulation of *reg2bit.bde* is shown in Fig, 19.5. Note that when the *load* signal is high the two bits *inp*(1:0) get latched into the register outputs *q*(1:0) on the next rising edge of the clock.

To test this 2-bit register on the FPGA board you can create the top-level design shown in Fig. 19.6 and download the resulting bit file to the FPGA board. Pressing *btn*(2) will display the switch setting *sw*(1:0) on the two rightmost LEDs. Note that this clock is running at 50 MHz but switch values get stored in the register only when the load signal is high.

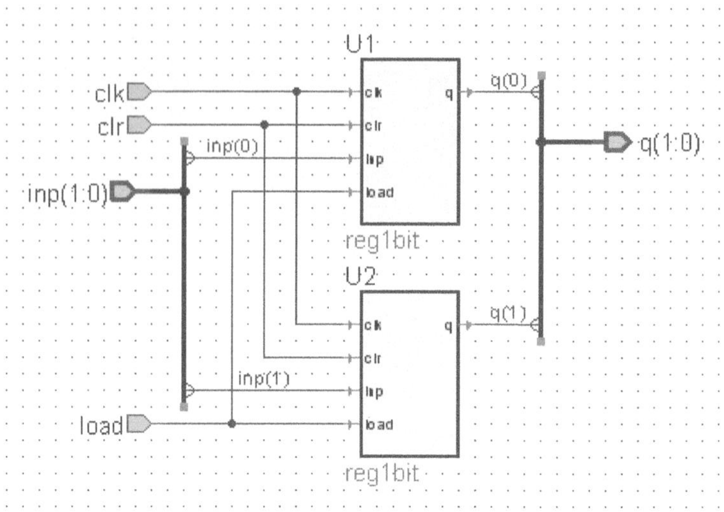

Figure 19.4 A 2-bit register *reg2bit.bde*

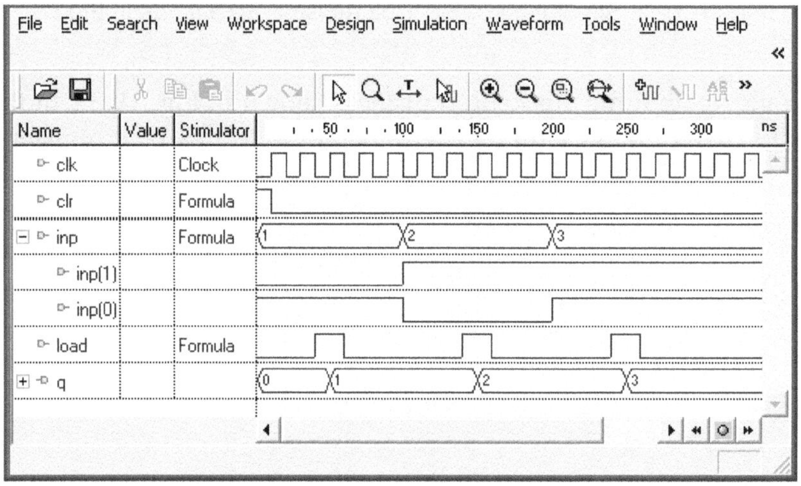

Figure 19.5 Simulation of the 2-bit register *reg2bit.bde* shown in Fig. 19.4

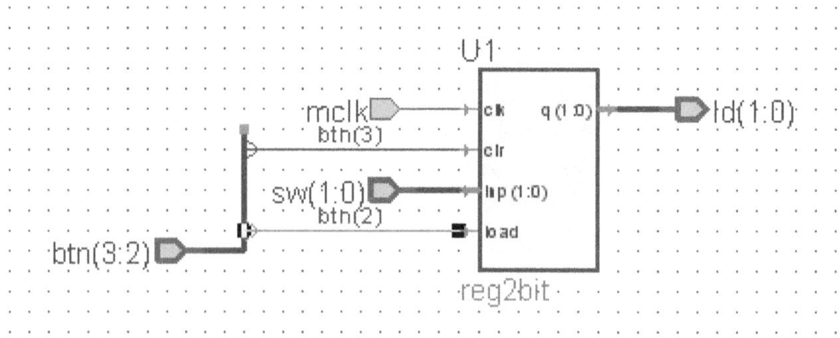

Figure 19.6 Top-level design *reg2bit_top.bde* for testing *reg2bit*

Example 20

N-Bit Register in VHDL

In this example we will show how to make an *N*-bit register in VHDL and test it on the FPGA board.

Prerequisite knowledge:
Example 17 – D Flip-Flops in VHDL
Example 19 – Registers

20.1 *N*-Bit Register

In Example 19 we saw how to make a 2-bit register using two 1-bit registers. To make an N-bit register it is easier to generalize the VHDL program we used in Example 17 to make a D flip-flop. For an *N*-bit register, if *load* is 1 then on the next rising edge of the clock *clk* the *N*-bit input *d*(*N*-1:0) is latched to the *N*-bit output *q*(*N*-1:0). A VHDL program for this register is shown in Listing 20.1. A simulation of this program for $N = 8$ is shown in Fig. 20.1.

Listing 20.1 register.vhd

```
-- Example 20: An N-bit register
library IEEE;
use IEEE.STD_LOGIC_1164.all;

entity reg is
       generic(N:integer := 8);
       port(
             load : in STD_LOGIC;
             clk : in STD_LOGIC;
             clr : in STD_LOGIC;
             d : in STD_LOGIC_VECTOR(N-1 downto 0);
             q : out STD_LOGIC_VECTOR(N-1 downto 0)
          );
end register;

architecture reg of reg is
begin
      process(clk, clr)
      begin
            if clr = '1' then
                  q <= (others => '0');
            elsif clk'event and clk = '1' then
                  if load = '1' then
                        q <= D;
                  end if;
      end if;
      end process;
end reg;
```

Note the use of the *generic* statement in Listing 20.1, which defines the bus width *N* to have a default value of 8. We can override this value and make it 4 by right-clicking the *register* symbol in the top-level design in Fig. 20.2, selecting *Properties*, clicking on the *Generics* tab, and changing the actual value of *N* to 4.

Generate the top-level design using the BDE in Active-HDL, implement the design, and download the bit file to the FPGA board. When you press *btn*(2) the switch settings will be loaded into the register and the register contents will be displayed on the 7-segment displays. Note that although the register clock is the 50 MHz *mclk*, data will be loaded into the register only when the load signal is high.

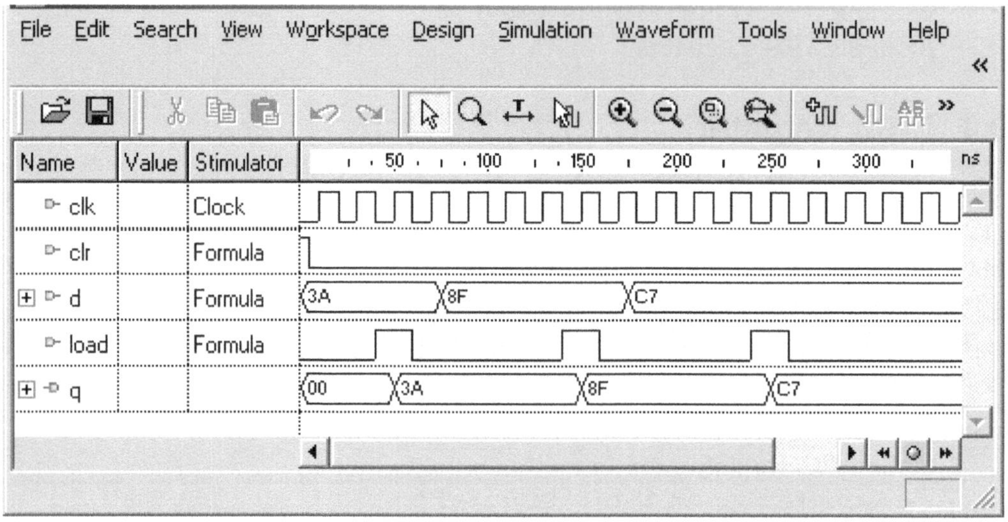

Figure 20.1 Simulation of the VHDL program in Listing 20.1

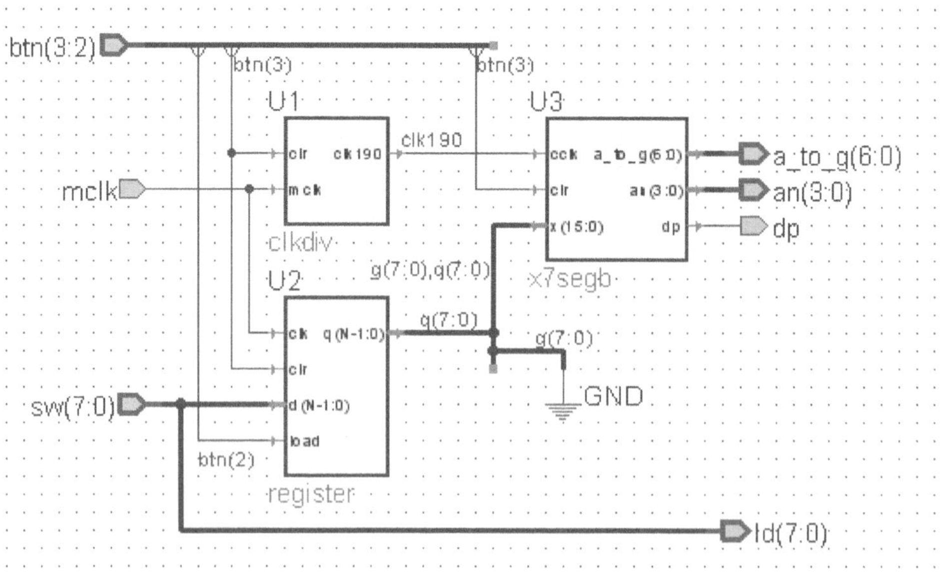

Figure 20.2 Top-level design *register_top.bde* for testing *register.vhd*

Example 21

Shift Registers

In this example we will show how to make a 4-bit and *N*-bit shift register in VHDL.

Prerequisite knowledge:
Example 17 – D Flip-Flops in VHDL
Example 19 – Registers

21.1 4-Bit Shift Register

An *N*-bit shift register contains *N* flip-flops. At each clock pulse data are shifted from one flip-flop to the next. A 4-bit shift register called *shift4.bde* is shown in Fig. 21.1. Serial data in the form of a string of bits is fed into the left-most flip-flop via *data_in*. At each clock pulse whatever is at *data_in* is moved to *q*(3), the old value at *q*(3) goes to *q*(2), the old value at *q*(2) goes to *q*(1), and the old value at *q*(1) goes to *q*(0). Note that all data values are shifted simultaneously on the same rising edge of the clock. A simulation of this shift register is shown in Fig. 21.2.

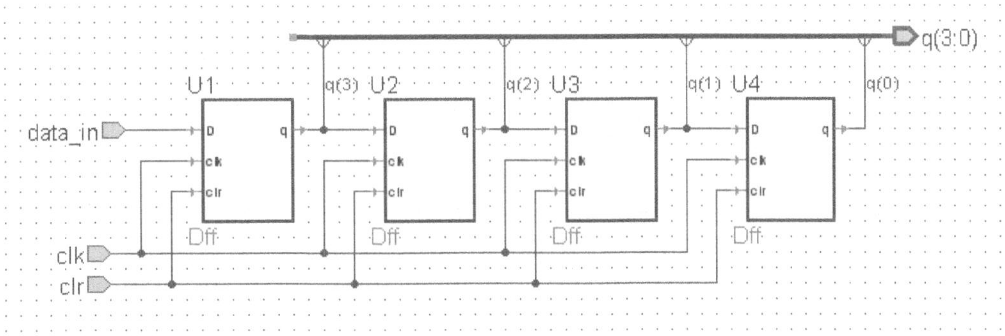

Figure 21.1 A 4-bit shift register *shift4.bde*

The 4-bit shift register shown in Fig. 21.1 can be modeled using the VHDL program shown in Listing 21.1. In the shift register program shown, we used the two signal assignment statements:

```
qs(3) <= data_in;
qs(2 downto 0) <= qs(3 downto 1);
```

These two statements are equivalent to

```
q(3) <= data_in;
q(2) <= q(3);
q(1) <= q(2);
q(0) <= q(1);
```

Note that the way we want the shift register to work is that the new value of $q(2)$ gets the old value of $q(3)$; i.e. the value that $q(3)$ had at the beginning of the process, and *not* the new value of *data_in* that $q(3)$ gets assigned within the process. This is exactly what the signal assignment operator <= will do for us. Had we used variables and the variable assignment operator := instead we would not have gotten a shift register at all, but just a 4-bit register in which all outputs get assigned *data_in* on the rising edge of the clock. Note that the statement

```
q <= qs;
```

is located outside of the process. This is important because *qs* is a signal that doesn't get updated until the end of the process. If we had assigned *q* its value inside the process it would also be updated at the end of the process, but with the previous value of *qs*, which would be wrong. By putting the *q* assignment statement outside the process it is executed concurrently with the process. This VHDL program will give the same simulation as shown in Fig. 21.2.

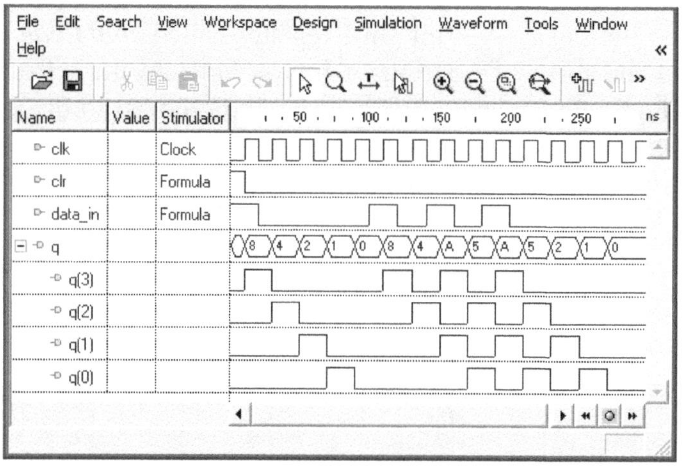

Figure 21.2 Simulation of the 4-bit shift register *shift4.bde*

Listing 21.1 shift4.vhd

```
-- Example 21: 4-bit shift register
library IEEE;
use IEEE.STD_LOGIC_1164.all;

entity shift4 is
        port(
              clk : in STD_LOGIC;
              clr : in STD_LOGIC;
              data_in : in STD_LOGIC;
              q : out STD_LOGIC_VECTOR(3 downto 0)
            );
end shift4;
```

Listing 21.1 (cont.) shift4.vhd

```
architecture shift4 of shift4 is
signal qs: STD_LOGIC_VECTOR(2 downto 0);

begin
	-- 4-bit shift register
	process(clk, clr)
	begin
		if clr = '1' then
			qs <= "0000";
		elsif clk'event and clk = '1' then
			qs(3) <= data_in;
			qs(2 downto 0) <= qs(3 downto 1);
		end if;
	end process;
	q <= qs;
end shift4;
```

Listing 21.2 is a generalization of Listing 21.1 that will produce an *N*-bit shift register.

Listing 21.2 ShiftReg.vhd

```
-- Example 21: N-bit shift register
library IEEE;
use IEEE.STD_LOGIC_1164.all;

entity ShiftReg is
	generic(N:integer := 8);
	port(
		clk : in STD_LOGIC;
		clr : in STD_LOGIC;
		data_in : in STD_LOGIC;
		q : out STD_LOGIC_VECTOR(N-1 downto 0)
	    );
end ShiftReg;

architecture ShiftReg of ShiftReg is
signal qs: STD_LOGIC_VECTOR(N-1 downto 0);
begin
	-- N-bit shift register
	process(clk, clr)
	begin
		if clr = '1' then
			qs <= (others => '0');
		elsif clk'event and clk = '1' then
			qs(N-1) <= data_in;
			qs(N-2 downto 0) <= qs(N-1 downto 1);
		end if;
	end process;
	q <= qs;
end ShiftReg;
```

Example 22

Ring Counters

In this example we will show how to make an *N*-bit ring counter in VHDL and test it on the FPGA board.

Prerequisite knowledge:
Example 17 – D Flip-Flops in VHDL
Example 21 – Shift Registers

22.1 4-Bit Ring Counter

If the output *q*(0) in the shift register in Fig. 21.1 is connected back to the input of the *q*(3) flip-flop, and if only a single 1 is present in the four flip-flops, we have what is called a *ring counter* as shown in Fig. 22.1. We will initially set *q*(0) to 1 by connecting the *clr* signal to the *set* input of flip-flop *q*(0) rather than to the *clr* input. The single 1 in this ring counter is continuously cycled around all four flip-flops. This means that the output *q*(*i*) on each flip-flop will go *HI* once every four clock cycles — but this pulse will be out of phase by one clock cycle from one flip-flop to the next. We will therefore have generated a 4-phase clock that can be used as the basis for various timing circuits.

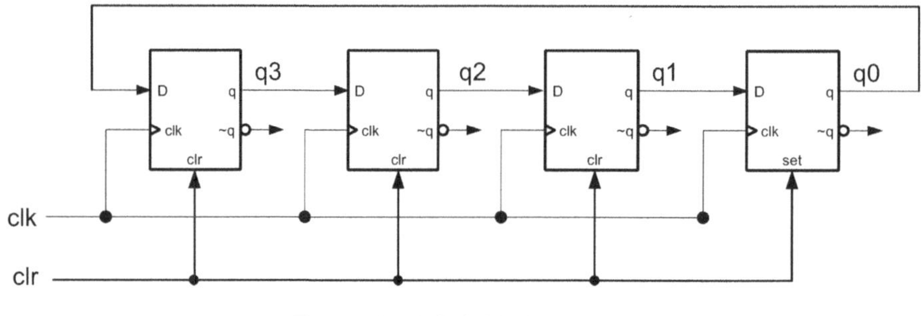

Figure 22.1 A 4-bit ring counter

22.2 *N*-Bit Ring Counter

A VHDL program for an *N*-bit ring counter is given in Listing 22.1. Note that when *clr* is equal to 1 the value of *q* is set to 1; i.e. *q*(0) is 1 and all other values of *q*(*i*) are zeros. The simulation of this VHDL program for the default value of *N* = 4 is shown in Fig. 22.2. Note how the four signals *q*(3:0) form a 4-phase clock at ¼ the frequency of the input clock, *clk*.

To test the ring counter on the FPGA board the top-level design *ring8_top.bde* shown in Fig. 22.3 will cause a single bit to cycle around all eight LEDs. You can set the value of the generic parameter *N* to 8 by right-clicking the *ring* symbol in the top-level design in Fig. 22.3, selecting *Properties*, clicking on the *Generics* tab, and changing the actual value of *N* to 8.

Listing 22.1 ring.vhd

```
--Example 22: N-bit ring counter
library IEEE;
use IEEE.STD_LOGIC_1164.all;

entity ring is
       generic(N:integer := 8);
       port(
             clk : in STD_LOGIC;
             clr : in STD_LOGIC;
             q : out STD_LOGIC_VECTOR(N-1 downto 0)
           );
end ring;

architecture ring of ring is
signal qs: STD_LOGIC_VECTOR(N-1 downto 0);
begin
      -- 4-bit ring counter
      process(clk, clr)
      begin
            if clr = '1' then
                  qs <= (others => '0');
                  qs(0) <= '1';
            elsif clk'event and clk = '1' then
                  qs(N-1) <= qs(0);
                  qs(N-2 downto 0) <= qs(N-1 downto 1);
            end if;
      end process;
      q <= qs;
end ring;
```

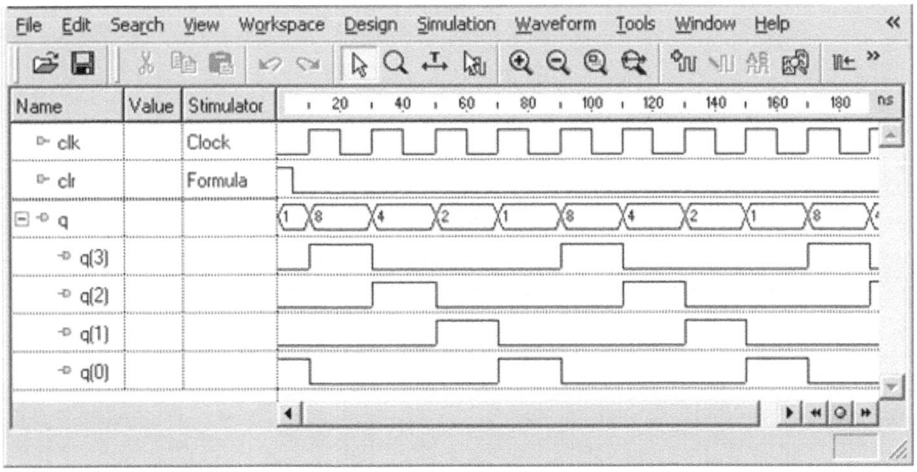

Figure 22.2 Simulation of the VHDL program in Listing 22.1

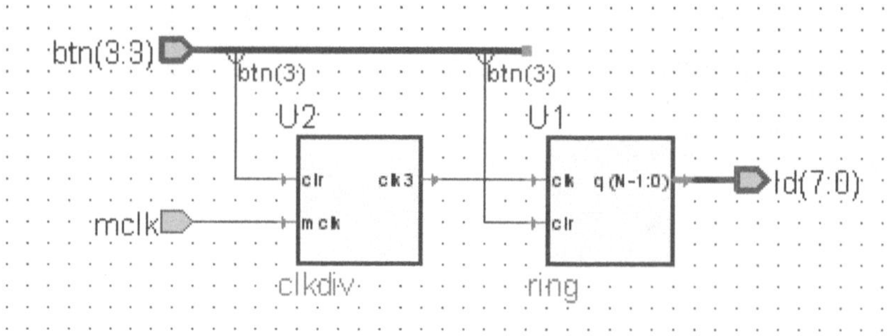

Figure 22.3 Top-level design *ring8_top.bde* for testing *ring.vhd*

Example 23

Johnson Counters

In this example we will show how to make Johnson counter in VHDL and test it on the FPGA board.

Prerequisite knowledge:
Example 17 – D Flip-Flops in VHDL
Example 21 – Shift Registers

23.1 *N*-Bit Johnson Counter

A 4-bit *Johnson counter* is shown in Fig. 23.1. The *clr* signal will cause all *q*(*i*) outputs to be zero. The value of ~*q*(0) is fed into the D input of *q*(3). A VHDL program for an *N*-bit Johnson counter with a default value of *N* = 4 is given in Listing 23.1. A simulation of the 4-bit Johnson counter is shown in Fig. 23.2.

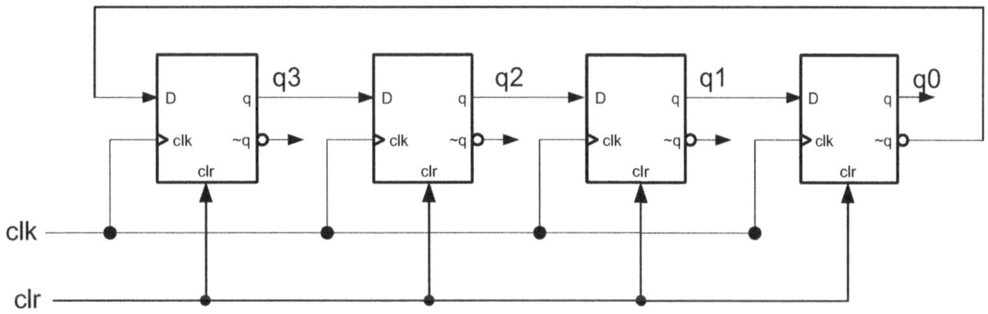

Figure 23.1 A 4-bit Johnson counter

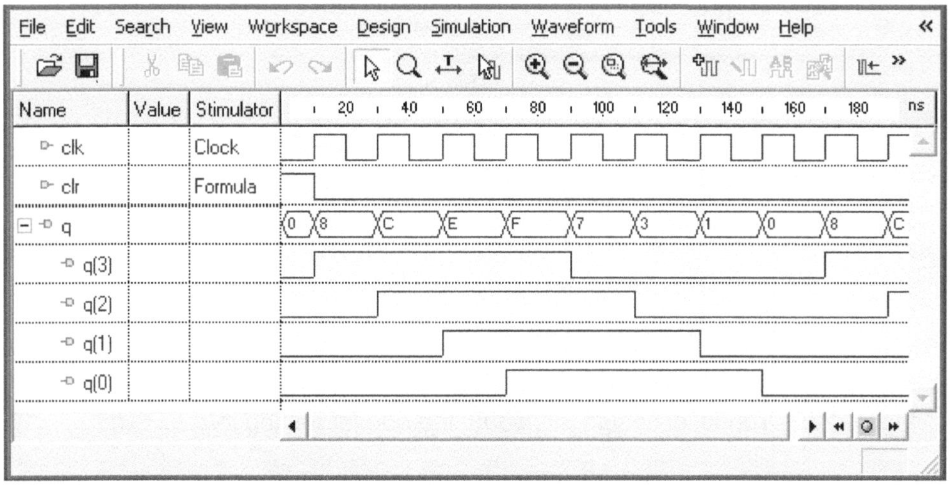

Figure 23.2 Simulation of the VHDL program in Listing 23.1

Listing 23.1 johnson.vhd

```
-- Example 23: N-bit Johnson counter
library IEEE;
use IEEE.STD_LOGIC_1164.all;

entity johnson is
	generic(N:integer := 8);
	port(
		clk : in STD_LOGIC;
		clr : in STD_LOGIC;
		q : out STD_LOGIC_VECTOR(N-1 downto 0)
	    );
end johnson;

architecture johnson of johnson is
signal qs: STD_LOGIC_VECTOR(N-1 downto 0);
begin
	-- N-bit Johnson counter
	process(clk, clr)
	begin
		if clr = '1' then
			qs <= (others => '0');
		elsif clk'event and clk = '1' then
			qs(N-1) <= not qs(0);
			qs(N-2 downto 0) <= qs(N-1 downto 1);
		end if;
	end process;
	q <= qs;
end johnson;
```

To test the Johnson counter on the FPGA board the top-level design *johnson8_top.bde* shown in Fig. 23.3 will cause the eight LEDs to display the output of an 8-bit Johnson counter. Remember to set the value of the generic parameter *N* to 8 by right-clicking the *johnson* symbol in the top-level design in Fig. 23.3, selecting *Properties*, clicking on the *Generics* tab, and changing the actual value of *N* to 8.

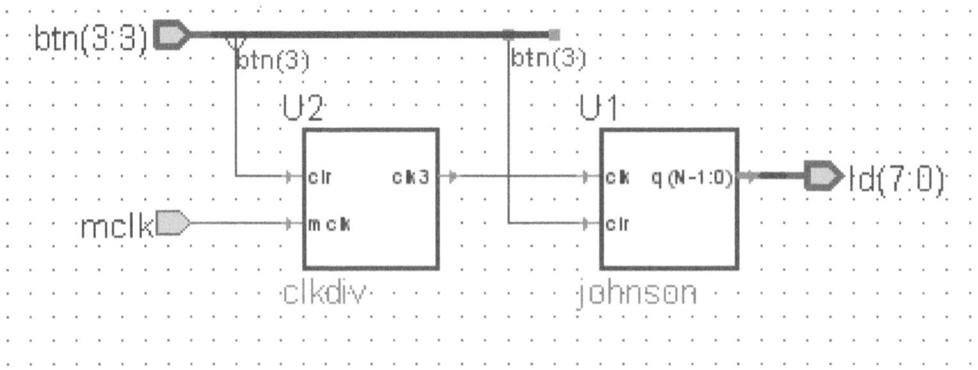

Figure 23.3 Top-level design *johnson8_top.bde* for testing *johnson.vhd*

Example 24

Debounce Pushbuttons

In this example we will show how to debounce the four pushbuttons on the FPGA board.

Prerequisite knowledge:
Example 17 – D Flip-Flops in VHDL
Example 21 – Shift Registers

24.1 Debouncing the Pushbuttons

When you press any of the pushbuttons on the BASYS or Nexys-2 boards they may bounce slightly for a few milliseconds before settling down. This means that instead of the input to the FPGA going from 0 to 1 cleanly it may bounce back and forth between 0 and 1 for a few milliseconds. This can be a serious problem in sequential circuits where actions take place on the rising edge of a clock signal. Because this clock signal changes much faster than the switch bouncing it is possible for erroneous values to be latched into registers. For this reason it is necessary to debounce the pushbutton switches when you use them in sequential circuits.

The circuit shown in Fig. 24.1 can be used to debounce a pushbutton input signal, *inp*. The frequency of the input clock, *cclk*, must be low enough that the switch bouncing is over before three clock periods. We will normally use the same 190 Hz frequency for *cclk* that is used to refresh the 7-segment displays.

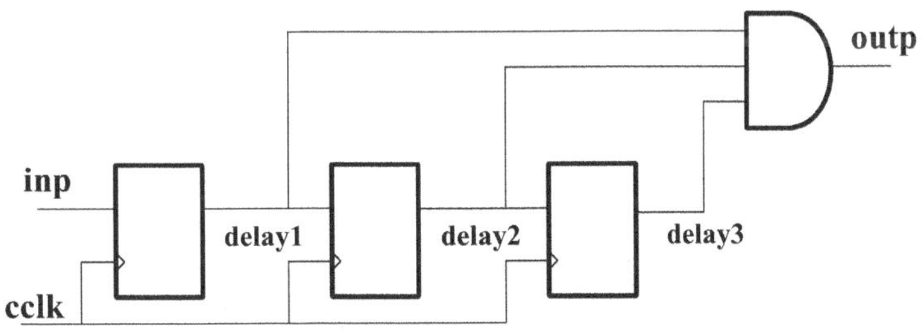

Figure 24.1 Debounce circuit

Listing 24.1 is a VHDL program that implements four versions of the debounce circuit in Fig. 24.1 – one for each of the four pushbuttons on the FPGA board. You can understand how this debounce circuit works by looking at the simulation shown in Fig. 24.2 where a bouncing input signal is used as a stimulator for *inp*(0). We have indicated bouncing on both pressing and releasing the button. Note that the resulting ouput signal *outp*(0) is a clean signal with all bouncing effects removed. The output will not go high until the input has been high for three clock cycles. The output will stay low as long as any bounces do not remain high for three clock cycles. Thus, it is important to use a low frequency for *cclk* to make sure all debounces are eliminated.

Listing 24.1 debounce4.vhd

```
-- Example 24: debounce 4 pushbuttons
library IEEE;
use IEEE.STD_LOGIC_1164.all;

entity debounce4 is
       port(
              inp : in STD_LOGIC_VECTOR(3 downto 0);
              cclk : in STD_LOGIC;
              clr : in STD_LOGIC;
              outp : out STD_LOGIC_VECTOR(3 downto 0)
            );
end debounce4;

architecture debounce4 of debounce4 is
signal delay1, delay2, delay3: STD_LOGIC_VECTOR(3 downto 0);
begin
      process(cclk, clr)
      begin
            if clr = '1' then
                  delay1 <= "0000";
                  delay2 <= "0000";
                  delay3 <= "0000";
            elsif cclk'event and cclk = '1' then
                  delay1 <= inp;
                  delay2 <= delay1;
                  delay3 <= delay2;
            end if;
      end process;
      outp <= delay1 and delay2 and delay3;
end debounce4;
```

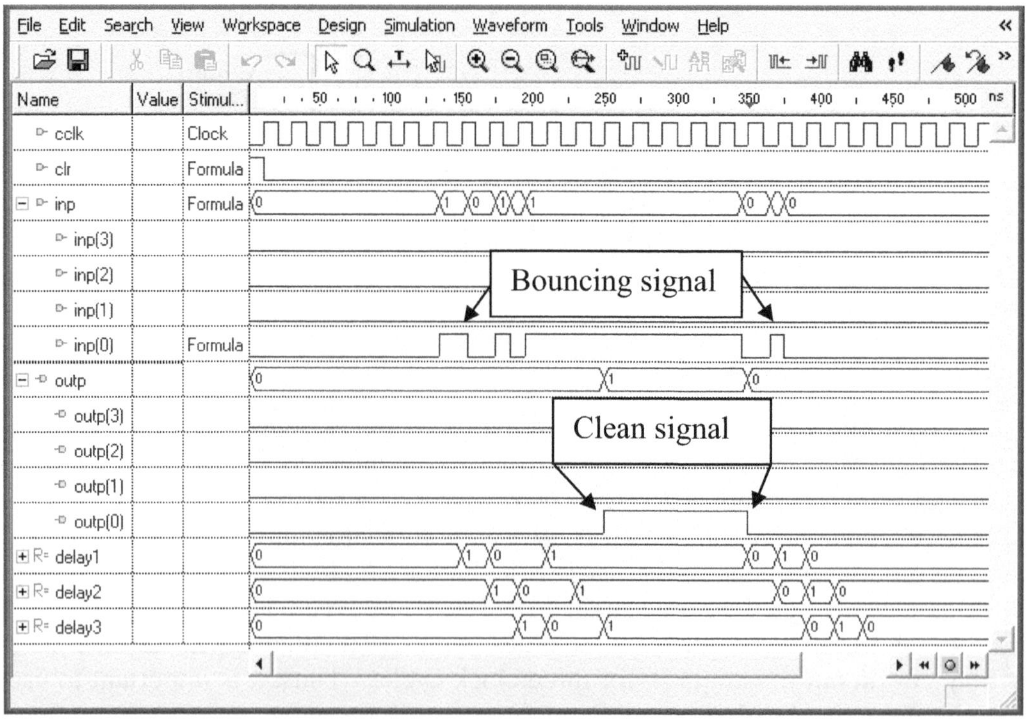

Figure 24.2 Simulation of the VHDL program in Listing 24.1

Example 25

Clock Pulse

In this example we will show how to generate a single clock pulse when a pushbutton is pressed.

Prerequisite knowledge:
Example 21 – Shift Registers
Example 24 – Debounce Pushbuttons

25.1 Generating a Single Clock Pulse

A very useful circuit that will produce a single clean clock pulse is shown in Fig. 25.1. The only difference from the debounce circuit in Fig. 24.1 is that the complement of *delay3* is the last input to the AND gate. Listing 25.1 is a VHDL program for this clock pulse circuit and its simulation is shown in Fig. 25.2.

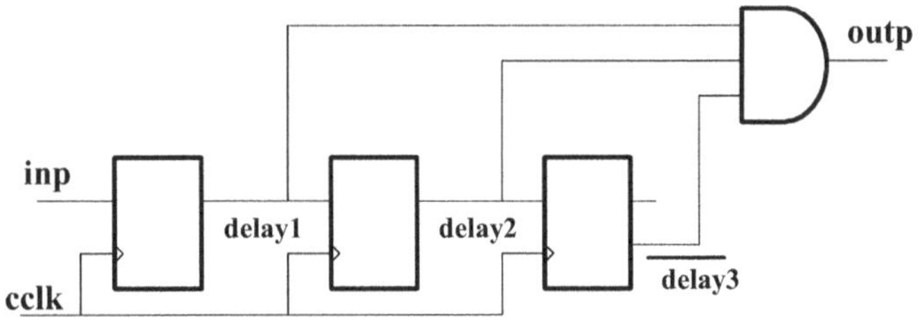

Figure 25.1 Clock pulse circuit

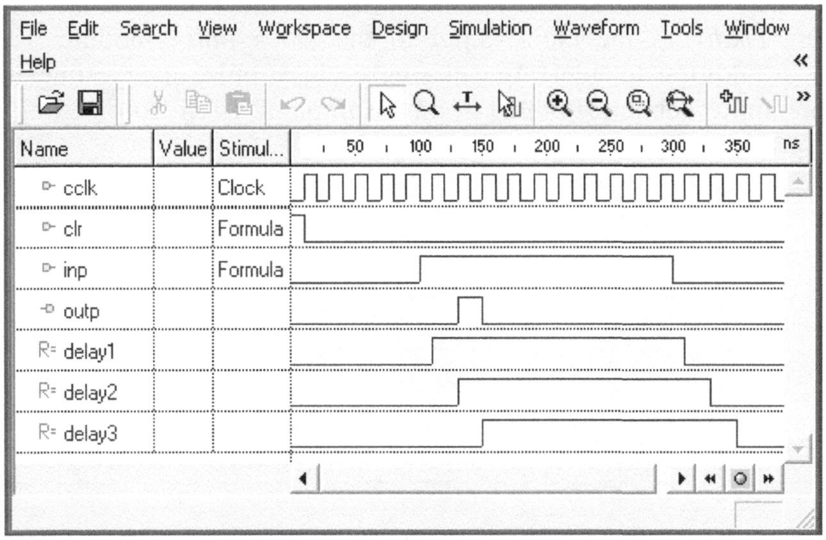

Figure 25.2 Simulation of the VHDL program in Listing 25.1

Listing 25.1 clock_pulse.vhd

```
-- Example 25: Clock Pulse
library IEEE;
use IEEE.STD_LOGIC_1164.all;

entity clock_pulse is
       port(
             inp : in STD_LOGIC;
             cclk : in STD_LOGIC;
             clr : in STD_LOGIC;
             outp : out STD_LOGIC
           );
end clock_pulse;

architecture clock_pulse of clock_pulse is
signal delay1, delay2, delay3: STD_LOGIC;
begin
      process(cclk, clr)
      begin
            if clr = '1' then
                  delay1 <= '0';
                  delay2 <= '0';
                  delay3 <= '0';
            elsif cclk'event and cclk = '1' then
                  delay1 <= inp;
                  delay2 <= delay1;
                  delay3 <= delay2;
            end if;
      end process;
      outp <= delay1 and delay2 and not delay3;
end clock_pulse;
```

In the top-level design *clock_pulse_top.bde* shown in Fig. 25.3 we use the clock pulse module to generate the clock input to a 16-bit counter from Example 8. When you press *btn*(2) the 7-segment display will increment by 1. Note that we use the 190 Hz clock output from *clkdiv* as the clock input to the *clock_pulse* module. This will debounce *btn*(2) in addition to generating a single clock pulse every time you press *btn*(2). It is important to debounce *btn*(2) in this example to keep from getting multiple counts when you press *btn*(2).

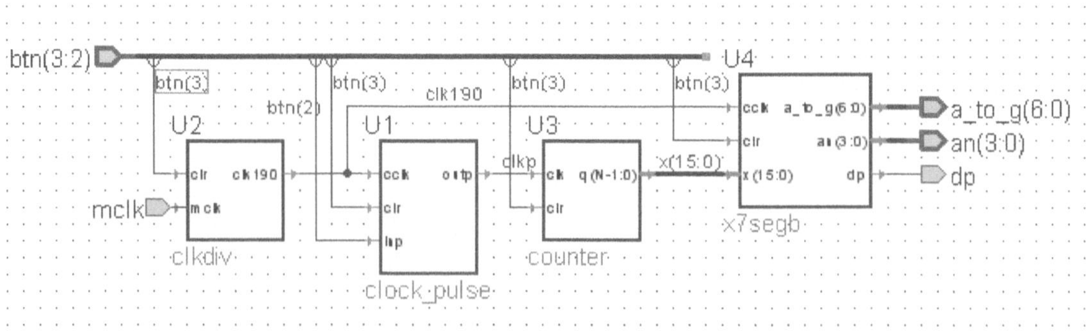

Figure 25.3 Top-level design *clock_pulse_top.bde* for testing *clock_pulse.vhd*

Example 26

Arbitrary Waveform

In this example we will show how to generate an arbitrary waveform that repeats itself every 16 clock cycles.

Prerequisite knowledge:
Example 8 – Counters
Appendix D – Karnaugh Maps

26.1 Generating the Morse Code for EAT

You own a restaurant and want to put a sign out in front that says EAT. But you only have a single light bulb. So you decide to blink the word EAT in Morse code! The Morse code for E is dot, for A is dot-dash, and for T is dash. Therefore you need to blink dot: dot-dash: dash in Morse code.

We can use the output of a 4-bit counter as the input to a combinational logic circuit with output *eat*. For a dot the output will be high for one clock cycle. For a dash the output will be high for three clock cycles. The truth table to generate the Morse code for EAT is shown in Fig. 26.1.

q(3)	q(2)	q(1)	q(0)	eat	
0	0	0	0	0	
0	0	0	1	0	
0	0	1	0	0	
0	0	1	1	1	} E
0	1	0	0	0	
0	1	0	1	0	
0	1	1	0	1	
0	1	1	1	0	
1	0	0	0	1	} A
1	0	0	1	1	
1	0	1	0	1	
1	0	1	1	0	
1	1	0	0	0	
1	1	0	1	1	
1	1	1	0	1	} T
1	1	1	1	1	

Figure 26.1 Truth table for generating the Morse code for EAT

The logic equation for *eat* can be found by using the K-map shown in Fig. 26.2. The resulting logic equation is shown in the VHDL program *eat.vhd* shown in Listing 26.1. Figure 26.2 shows a top-level design that will blink the Morse code for EAT on *ld*(0) on the FPGA board. Try it.

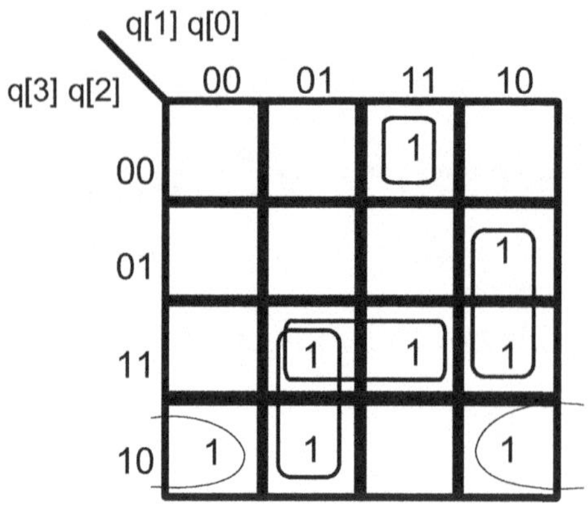

Figure 26.2 Karnaugh map for the truth table in Fig. 26.1

Listing 26.1 eat.vhd

```
-- Example 26: Morse code for EAT
library IEEE;
use IEEE.STD_LOGIC_1164.all;

entity eat is
       port(
              q : in STD_LOGIC_VECTOR(3 downto 0);
              eat : out STD_LOGIC
           );
end eat;

architecture eat of eat is
begin
      eat <= (not q(3) and not q(2) and q(1) and q(0))
              or (q(2) and q(1) and not q(0))
              or (q(3) and q(2) and q(0))
              or (q(3) and not q(1) and q(0))
              or (q(3) and not q(2) and not q(0));

end eat;
```

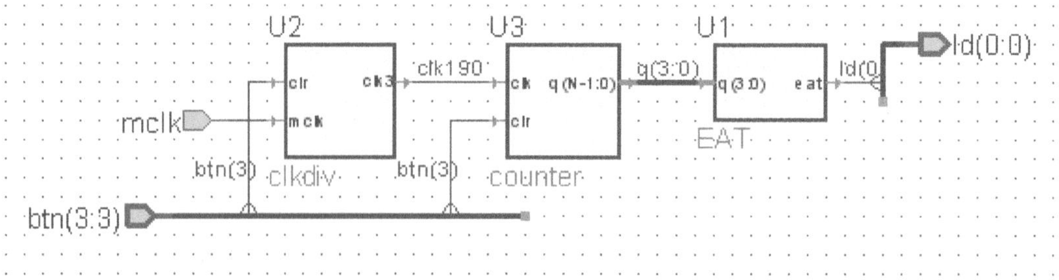

Figure 26.3 Top-level design *eat_top.bde* for generating the Morse code for EAT

Example 27

Pulse-Width Modulation (PWM)

In this example we will show how to generate a PWM signal that can be used to control the speed of a DC motor.

Prerequisite knowledge:
Example 8 – Counters
Example 15 – Comparators

27.1 Controlling the Speed of a DC Motor

When connecting a motor or some other load that may draw significant current to a digital circuit such as a CPLD, FPGA, or microcontroller a separate circuit and power supply are needed. The speed of a DC motor depends on the voltage applied to the motor – the higher the voltage the faster the motor will turn. The polarity of the voltage connected to the motor will determine which way the motor turns. If you want to be able to change the direction of the motor using a digital circuit you will need to use an H-bridge. Digilent sells two Pmod H-bridge modules that make it easy to plug into the 6-pin connectors on the BASYS or Nexys-2 boards.

The basic idea of how an H-bridge works can be seen from the diagram shown in Fig. 27.1. We show four relays in the legs of the H with the motor connected across the center of the H. The two top relays A and B are normally closed (where a high signal will open the relay) and the two bottom relays C and D are normally open (where a high signal will close the relay). In a real H-bridge these relays will normally be replaced with MOSFET transistors.

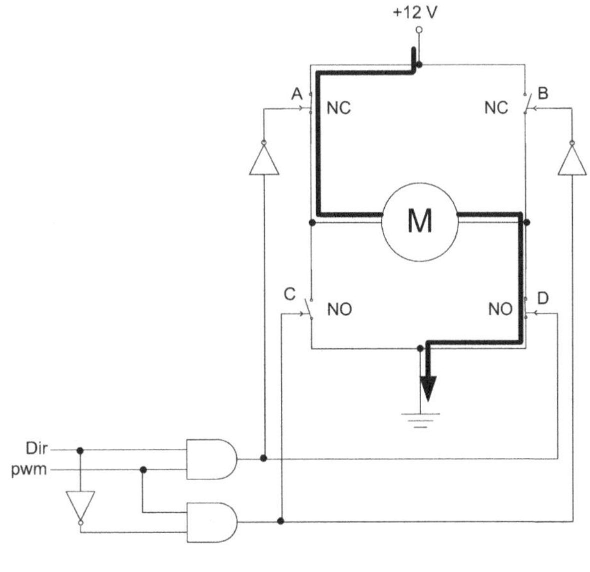

Figure 27.1 An H-bridge circuit can control the speed and direction of a DC motor

In Fig. 27.1 it is assumed that the two digital inputs *Dir* and *pwm* are both high. In this case relays A and D are both closed and relays B and C are both open. Current therefore flows through the motor from left to right in the figure as shown. This will cause the motor to rotate, say, clockwise. Assume now that the *Dir* input is brought low with *pwm* remaining high. This will cause relays B and C to close and relays A and D to open. Current will now flow through relays B and C and go through the motor from right to left in the figure. This will cause the motor rotate counterclockwise. Thus, the *Dir* input controls the direction of the motor.

To control the speed of a DC motor using a digital circuit one normally uses a pulse-width modulated signal of the type shown in Fig. 27.2. The period of this pulse train remains constant and the width of the high time, called *duty* in Fig. 27.2, is varied. The *duty cycle* of a PWM signal is defined as the percent time that the signal is high. That is,

$$\text{duty cycle} = \frac{duty}{period} \times 100\%$$

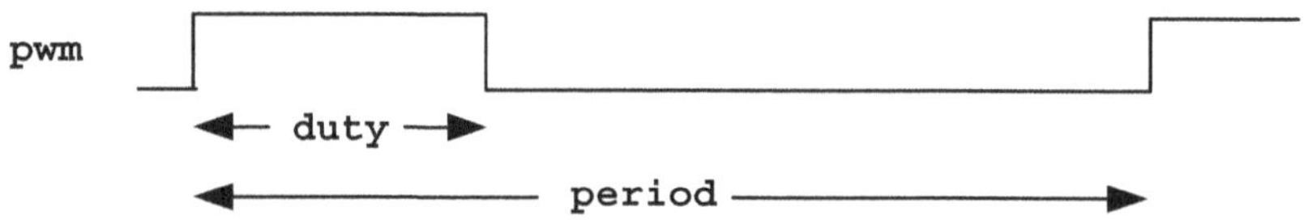

Figure 27.2 A pulse-width modulated signal

The average DC value of the *pwm* signal in Fig. 27.2 will be proportional to the duty cycle. A duty cycle of 100% will have a DC value equal to the maximum value of the *pwm* signal. A duty cycle of 50% will have a DC value equal to half of the maximum value of the *pwm* signal, and so forth. If the voltage across the motor is proportional to this *pwm* signal, then simply changing the pulse width *duty* and therefore the duty cycle changes the speed of the motor.

If this *pwm* signal is connected to the *pwm* input in Fig. 27.1 then when the *pwm* signal is high the motor is connected to one leg of the H-bridge. However, when the *pwm* input in low then all relays are open the it is as if we removed the motor from its power source. Thus, the average DC voltage across the motor will depend on the duty cycle of the *pwm* signal where increasing the duty cycle of the *pwm* signal will increase the speed of the motor.

27.2 Generating a PWM Signal

In this section we will show how to generate a pulse-width modulated (PWM) signal of the type shown in Fig. 27.2 using VHDL. To show how you might generate such a *pwm* signal consider the simulation shown in Fig. 27.3. The basic idea is to use a counter (a 4-bit counter is used for illustration in Fig. 27.2) and have the *pwm* signal go high if the *count* is less than *duty* and goes low otherwise. The counter will be reset to zero when the value of *count* is equal to $period - 1$. In the simulation shown in Fig. 27.3

the value of duty is 4 and the value of period is 13 (or hex D). The VHDL program that produced the simulation in Fig. 27.3 is given in Listing 27.1.

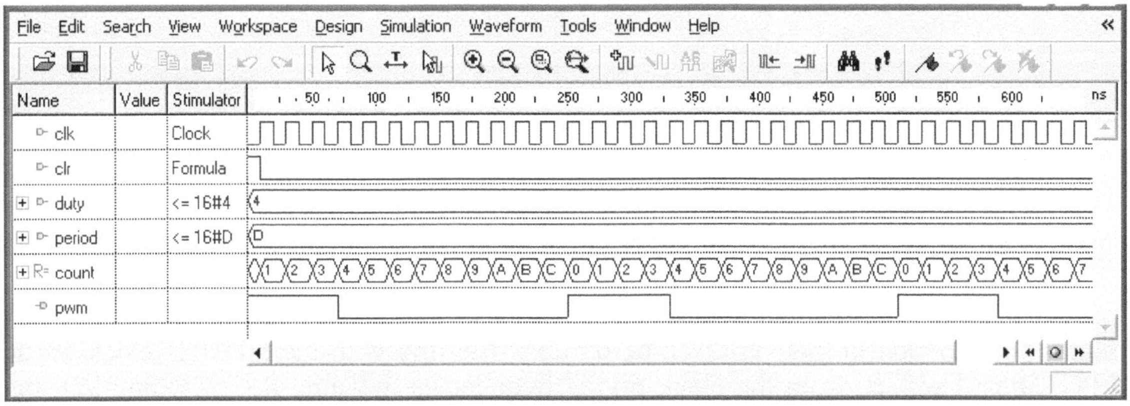

Figure 27.3 Simulation of a PWM signal

Listing 27.1 pwmg.vhd

```
-- Example 27: PWM module
library IEEE;
use IEEE.STD_LOGIC_1164.all;
use IEEE.STD_LOGIC_unsigned.all;

entity pwmg is
       generic( N : integer := 16);
       port(
              clr : in STD_LOGIC;
              clk : in STD_LOGIC;
              duty : in STD_LOGIC_VECTOR(N-1 downto 0);
              period : in STD_LOGIC_VECTOR(N-1 downto 0);
              pwm : out STD_LOGIC
            );
end pwmg;


architecture pwmg of pwmg is
signal count: STD_LOGIC_VECTOR(N-1 downto 0);
begin
      clk4: process(clk, clr)     -- 4-bit counter
      begin
             if clr = '1' then
                   count <= (others => '0');
             elsif clk'event and clk = '1' then
                   if count = period - 1 then
                          count <= (others => '0');
                   else
                          count <= count + 1;
                   end if;
             end if;
      end process clk4;
```

Listing 27.1 (cont.) pwmg.vhd

```
        pwmout: process(count)
        begin
           if count < duty then
              pwm <= '1';
           else
              pwm <= '0';
           end if;
        end process pwmout;
end pwmg;
```

27.3 Generating a 2 kHz PWM Signal

Suppose we wish to generate a 2 kHz PWM signal to control the speed of a motor. The period in this case will be 0.5 ms. Starting with the 50 MHz clock on the FPGA board we see from the clock divide frequencies in Table 8.1 of Example 8 that $q(14)$ has a period just over 0.5 ms (0.65536). We could therefore use an 8-bit counter driven by the clock frequency $q(6)$ (390.625 kHz) to control the PWM signal. This will allow us to set the period to 0.5 ms by choosing a value of *period* equal to

```
period = (0.5/0.65536)*255 = 195 (C3 hex)
```

The value of duty could then range from hex 00 to C3. The simulation shown in Fig. 27.4 uses a clock frequency of 390.625 kHz with an 8-bit counter ($N = 8$) and shows that this results in a 2 kHz PWM signal. If you were to use such a PWM signal to control the speed of a DC motor using a Pmod H-bridge from Digilent with the BASYS board you would output the *pwm* signal through one of the expansion connectors labeled JA, JB, JC, and JD. Four pins on each connector are connected to the FPGA pin numbers shown in Table 27.1. If you were to connect the H-bridge module to connector JA the *Dir* signal would be connected to JA1 from FPGA pin 81 and the *pwm* signal would be connected to JA2 from FPGA pin 91. You would need to add these two signals and pin numbers to the *basys2.ucf* file.

Table 27.1 Pin numbers for the BASYS board expansion connectors

JA				JB				JC				JD			
1	2	3	4	1	2	3	4	1	2	3	4	1	2	3	4
81	91	82	92	87	93	88	94	77	86	76	85	75	59	74	58

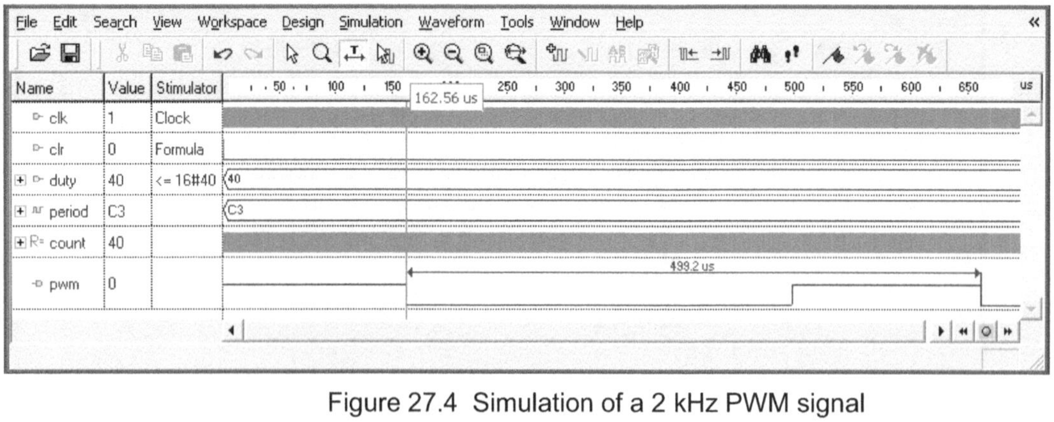

Figure 27.4 Simulation of a 2 kHz PWM signal

Example 28

Controlling the Position of a Servo

In this example we will show how to generate a PWM signal that can be used to control the position of a servo.

Prerequisite knowledge:
Example 8 – Counters
Example 27 – PWM

28.1 Controlling the Position of a Servo using PWM

A servo motor is a special type of device that contains a DC motor, some gears, a potentiometer, and electronic circuitry for position feedback control, all packaged in a single compact device. These servos are widely used in model airplanes and radio controlled cars and are therefore mass produced and very inexpensive. A typical servo of this type, the Futaba S3004, is shown in Fig. 28.1. This servo has three wires attached to it: the red wire goes to +5 volts, the black wire goes to ground, and the white wire goes to a PWM signal that controls the position of the motor shaft.

Figure 28.1
The Futaba S3004 servo

The motor shaft is prevented from moving more than about 90 degrees by limit stops. The PWM signal used to control the position of a servo is shown in Fig. 28.2. Note that the period is fixed at 20 ms and the pulse width varies from about 1.1 ms to 1.9 ms in order to move the shaft position through a total angle of about 90 degrees.

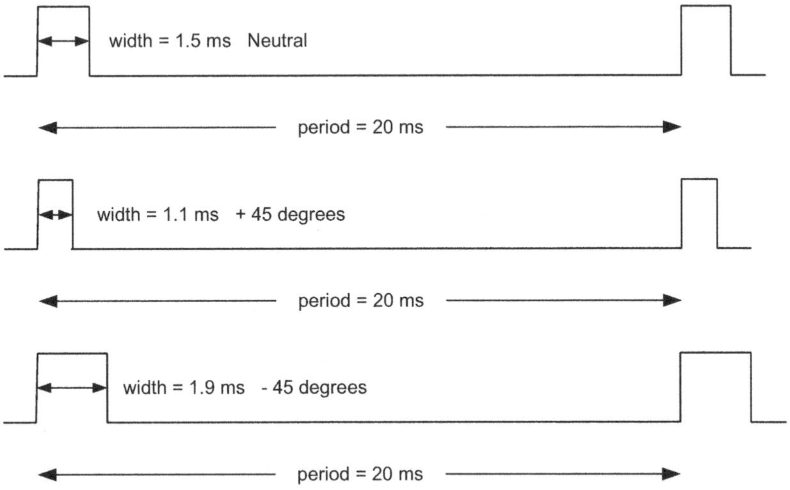

Figure 28.2 PWM signals for controlling the position of a servo

The VHDL program given in Listing 27.1 in Example 27 can be used to generate the PWM signal in Fig. 28.2. Starting with the 50 MHz clock on this board we see from the clock divide frequencies in Table 8.1 of Example 8 that *q*(19) has a period just over 20ms (20.97125). We could therefore use a 16-bit counter driven by the clock frequency *q*(3) (3.125 MHz) to control the PWM signal. This will allow us to set the period to 20 ms by choosing a value of *period* equal to

```
period = (20/20.97125)*65535 = 62500 (F424 hex)
```

In this case the value of *duty* in Fig. 27.2 would be about 4688 (1250 hex) for the neutral position, 3438 (D6E hex) for the +45 degree position, and 5938 (1732 hex) for the –45 degree position.

A simulation of the VHDL program in Listing 27.1 for $N = 16$ and a value of period equal to 62500 (F424 hex) and a value of *duty* equal to the neutral position of 4688 (1250 hex) is shown in Fig. 28.3. Note that the pulse width is 1.5 ms and the period is 20 ms.

Digilent sells a Pmod servo connector that makes it easy to connect a servo motor to the BASYS or Nexys-2 boards.

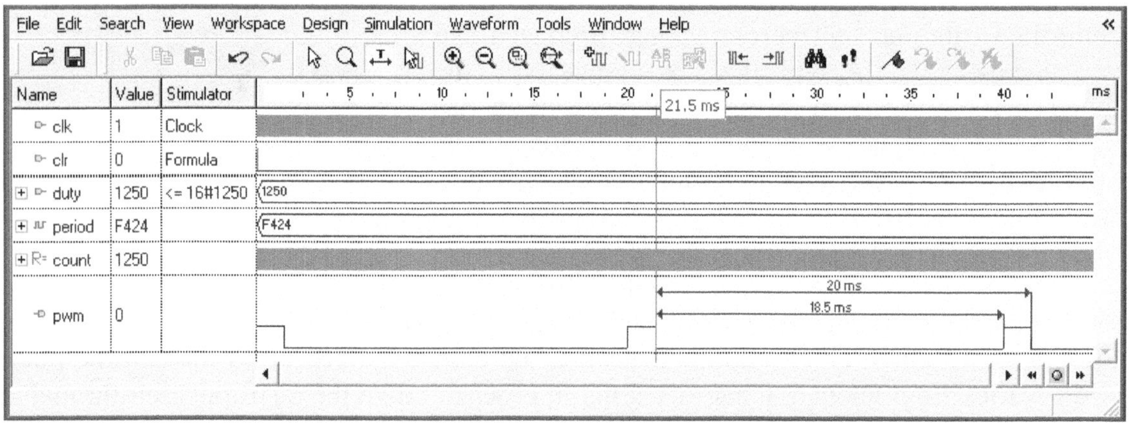

Figure 28.3 Simulation of a PWM signal for servo control

Example 29

Scrolling the 7-Segment Display

In this example we will show how to scroll a message on the 7-segment display. We will illustrate the process by scrolling the telephone number 248-656-1490 across the 7-segment displays.

Prerequisite knowledge:
Example 10 – 7-Segment Displays

29.1 Scrolling the 7-Segment Display

The basic idea in this example will be to modify our *x7seg* module by changing the *hex7seg* decoder part to display a dash (-) for a hex D input and to display a blank for a hex F input as shown in Fig. 29.1. The design shown in Fig. 29.2 shows how the scrolling is accomplished. The digits of the telephone number are stored in a 64-bit register *msg_array*(0:63). Fig. 29.2 shows how the sixteen bits *msg_array*(0:15) are connected to the input *x*(15:0) of the *x7seg_msg* module. On reset the 7-segment display will display the first four characters in the message, namely 248-. The bits in *msg_array*(0:63) are connected as a shift register so that we shift the contents of *msg_array*(0:63) four bits the the left on the rising edge of the clock with the value of *msg_array*(0:3) going into *msg_array*(60:63). Thus, after one clock pulse the value of *msg_array*(0:15) will contain 48-6. On each successive clock pulse the message will be shifted four bits to the left. We will use the 3 Hz clock to shift the characters on the 7-segment display.

```
-- Modified 7-segment decoder: hex7seg
   process(digit)
   begin
      case digit is
          when X"0" => a_to_g <= "0000001";
          when X"1" => a_to_g <= "1001111";
          when X"2" => a_to_g <= "0010010";
          when X"3" => a_to_g <= "0000110";
          when X"4" => a_to_g <= "1001100";
          when X"5" => a_to_g <= "0100100";
          when X"6" => a_to_g <= "0100000";
          when X"7" => a_to_g <= "0001101";
          when X"8" => a_to_g <= "0000000";
          when X"9" => a_to_g <= "0000100";
          when X"A" => a_to_g <= "0001000";
          when X"B" => a_to_g <= "1100000";
          when X"C" => a_to_g <= "0110001";
          when X"D" => a_to_g <= "1111110";      --dash
          when X"E" => a_to_g <= "0110000";
          when X"F" => a_to_g <= "1111111";      --blank
          when others => a_to_g <= "1111111";    --blank
        end case;
   end process;
```

Figure 28.1 Modified *hex7seg* module

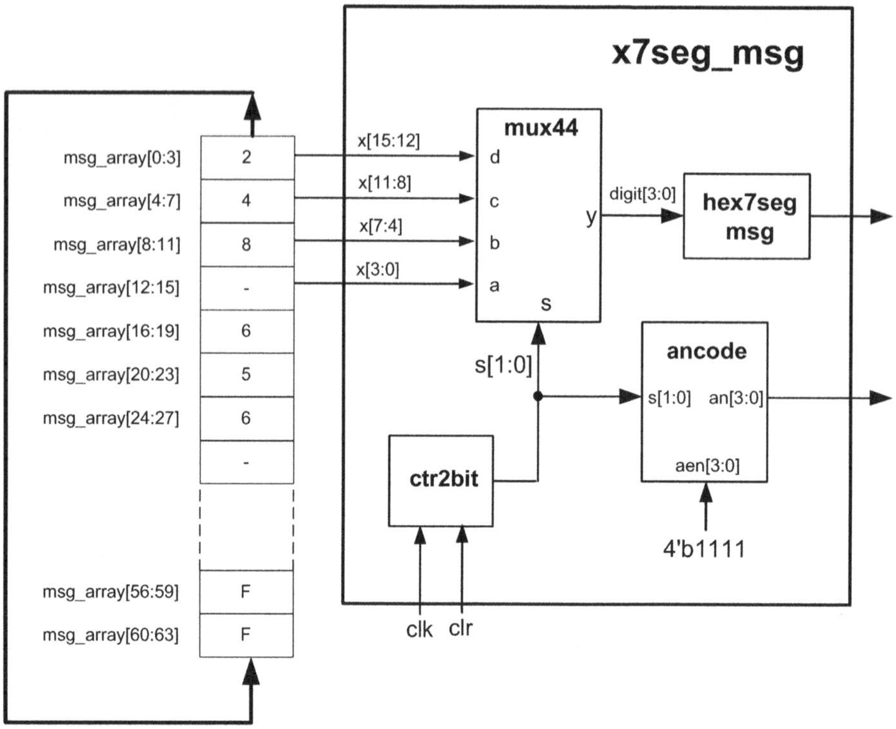

Figure 29.2 Scrolling a message on the 7-segment display

Listing 29.1 shows how to implement the shift register *msg_array*(0:63) in VHDL. Note that phone number is defined as a 64-bit constant using the *constant* statement

```
constant PHONE_NO: STD_LOGIC_VECTOR(63 downto 0) := X"248D656D1490FFFF";
```

The *process* in Listing 29.1 initializes the value of *msg_array* when the *clr* signal is asserted. Then on the rising edge of the clock the contents of *msg_array* are shifted four bits to the left in a circular fashion. The output *x*(15:0) of this shift array is just the value of *msg_array*(0:15). A simulation of this shift array is shown in Fig. 29.3.

The complete listing of the module *x7seg_msg* is shown in Listing 29.2 and the top_level design is shown in Listing 29.3. Implement and run this program on the FPGA board. Change the phone number to yours.

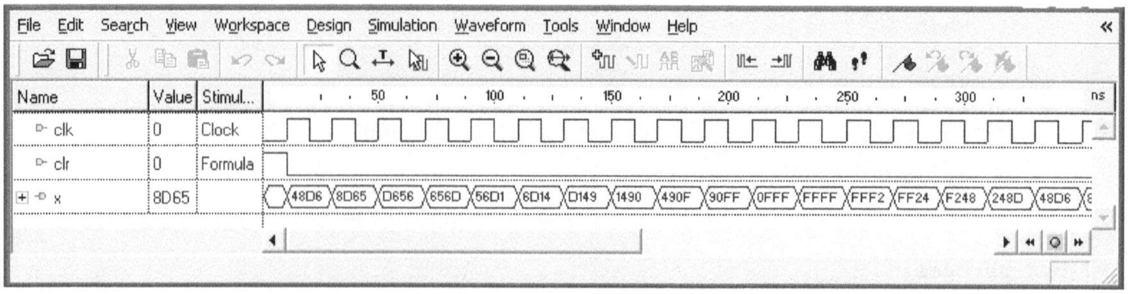

Figure 29.3 Simulation of the VHDL program in Listing 29.1

Listing 29.1 shift_array.vhd

```
-- Example 29a: shift_array
library IEEE;
use IEEE.STD_LOGIC_1164.all;

entity shift_array is
       port(
              clk : in STD_LOGIC;
              clr : in STD_LOGIC;
              x : out STD_LOGIC_VECTOR(15 downto 0)
              );
end shift_array;

architecture shift_array of shift_array is
signal msg_array: STD_LOGIC_VECTOR(0 to 63);
constant PHONE_NO: STD_LOGIC_VECTOR(63 downto 0):=X"248D656D1490FFFF";

begin
      process(clr, clk)
      begin
            if clr = '1' then
                  msg_array <= PHONE_NO;
            elsif (clk'event and clk = '1') then
                  msg_array(0 to 59) <= msg_array(4 to 63);
                  msg_array(60 to 63) <= msg_array(0 to 3);
            end if;
      end process;
      x <= msg_array(0 to 15);
end shift_array;
```

Listing 29.2 x7seg_msg.vhd

```
-- Example 29b: x7seg_msg - Display scrolling message
-- input cclk should be 190 Hz
library IEEE;
use IEEE.STD_LOGIC_1164.all;
use IEEE.STD_LOGIC_UNSIGNED.all;

entity x7seg_msg is
       port(
              x : in STD_LOGIC_VECTOR(15 downto 0);
              clk : in STD_LOGIC;
              clr : in STD_LOGIC;
              a_to_g : out STD_LOGIC_VECTOR(6 downto 0);
              an : out STD_LOGIC_VECTOR(3 downto 0);
              dp : out STD_LOGIC
            );
end x7seg_msg;

architecture x7seg_msg of x7seg_msg is
signal s: STD_LOGIC_VECTOR(1 downto 0);
signal digit: STD_LOGIC_VECTOR(3 downto 0);
signal aen: STD_LOGIC_VECTOR(3 downto 0);

begin
   aen <= "1111";
   dp <= '1';
```

Listing 29.2 (cont.) x7seg_msg.vhd

```
-- Quad 4-to-1 MUX: mux44
   process(s)
   begin
      case s is
           when "00" => digit <= x(3 downto 0);
           when "01" => digit <= x(7 downto 4);
           when "10" => digit <= x(11 downto 8);
           when others => digit <= x(15 downto 12);
      end case;
   end process;

-- 2-bit counter
   process(clk, clr)
   begin
      if clr = '1' then
         s <= "00";
      elsif clk'event and clk = '1' then
         s <= s + 1;
      end if;
   end process;

-- Digit select: ancode
   process(s, aen)
   begin
      an <= "1111";
      if aen(conv_integer(s)) = '1' then
         an(conv_integer(s)) <= '0';
      end if;
   end process;

-- 7-segment decoder: hex7seg
   process(digit)
   begin
      case digit is
          when X"0" => a_to_g <= "0000001";      --0
          when X"1" => a_to_g <= "1001111";      --1
          when X"2" => a_to_g <= "0010010";      --2
          when X"3" => a_to_g <= "0000110";      --3
          when X"4" => a_to_g <= "1001100";      --4
          when X"5" => a_to_g <= "0100100";      --5
          when X"6" => a_to_g <= "0100000";      --6
          when X"7" => a_to_g <= "0001101";      --7
          when X"8" => a_to_g <= "0000000";      --8
          when X"9" => a_to_g <= "0000100";      --9
          when X"A" => a_to_g <= "0001000";      --A
          when X"B" => a_to_g <= "1100000";      --b
          when X"C" => a_to_g <= "0110001";      --C
          when X"D" => a_to_g <= "1111110";      --dash
          when X"E" => a_to_g <= "0110000";      --E
          when X"F" => a_to_g <= "1111111";      --blank
          when others => a_to_g <= "1111111";    --blank
      end case;
   end process;

end x7seg_msg;
```

Listing 29.3 scroll_top.vhd

```
-- Example 29c: Scroll phone number on 7seg display
library IEEE;
use IEEE.STD_LOGIC_1164.all;

entity scroll_top is
       port(
              mclk : in STD_LOGIC;
              btn : in STD_LOGIC_VECTOR(3 downto 3);
              a_to_g : out STD_LOGIC_VECTOR(6 downto 0);
              an : out STD_LOGIC_VECTOR(3 downto 0);
              dp : out STD_LOGIC
            );
end scroll_top;

architecture scroll_top of scroll_top is
component clkdiv is
       port(
              mclk : in STD_LOGIC;
              clr : in STD_LOGIC;
              clk190 : out STD_LOGIC;
              clk3 : out STD_LOGIC
            );
end component;
component shift_array is
       port(
              clk : in STD_LOGIC;
              clr : in STD_LOGIC;
              x : out STD_LOGIC_VECTOR(15 downto 0)
            );
end component;
component x7seg_msg is
       port(
              x : in STD_LOGIC_VECTOR(15 downto 0);
              clk : in STD_LOGIC;
              clr : in STD_LOGIC;
              a_to_g : out STD_LOGIC_VECTOR(6 downto 0);
              an : out STD_LOGIC_VECTOR(3 downto 0);
              dp : out STD_LOGIC
            );
end component;
signal clr, clk190, clk3: STD_LOGIC;
signal x: STD_LOGIC_VECTOR(15 downto 0);
begin
  clr <= btn(3);

  U1: clkdiv
     port map (
        mclk => mclk, clr => clr, clk190 => clk190,
        clk3 => clk3);
  U2: shift_array
     port map (
        clk => clk3, clr => clr, x => x);
  U3: x7seg_msg
     port map (
        x => x, clk => clk190, clr => clr,
        a_to_g => a_to_g, an => an, dp => dp);

end scroll_top;
```

Example 30

Fibonacci Sequence

In this example we will show how to generate the Fibonacci sequence and display the result on the 7-segment display

Prerequisite knowledge:

Example 8 – Clocks and Counters
Example 10 – 7-Segment Displays
Example 19 – Registers

30.1 Fibonacci Sequence

The following sequence of numbers is known as the *Fibonacci sequence.*

(0, 1, 1, 2, 3, 5, 8, 13, 21, 34, ...)

The function for generating Fibonacci numbers is

$F(0) = 0$
$F(1) = 1$
$F(n + 2) = F(n) + F(n + 1)$ for all $n \geq 0$.

In other words, starting with a zero and a one, the next number is generated by adding the previous two numbers. This requires storing previous results, namely the two previous results. This can be done by using two registers *fn* and *fn1* as shown in Fig. 30.1. For each of these registers, if the load signal *rld* is 1, the value coming into the top gets latched to the output coming out the bottom on the rising edge of the clock input *clk*. The *clr* input signal causes an asynchronous resetting of the output to a predetermined initial value (0 for *fn* and 1 for *fn1*).

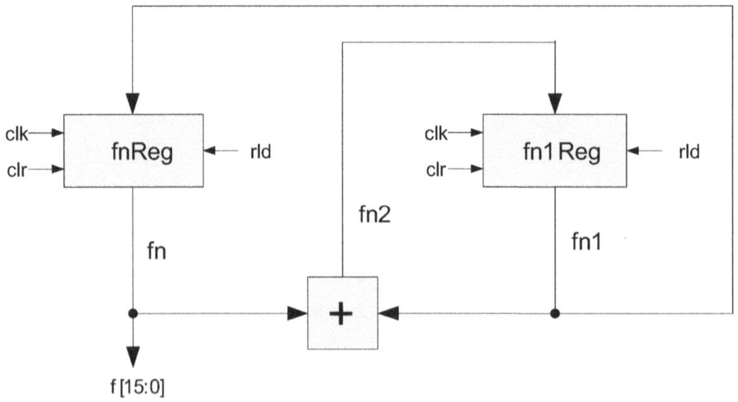

Figure 30.1 Circuit for computing the Fibonacci sequence

On the rising edge of the clock the value of *fn1* gets updated with the sum $fn2 = fn + fn1$ and *fn* gets updated with the old value of *fn1*. We will use 16-bit registers in the circuit in Fig. 30.1. It is possible to write a single VHDL program involving a single process that will implement the Fibonacci sequence algorithm as shown in Listing 30.1. A simulation of this VHDL program is shown in Fig. 30.2.

Listing 30.1 fib.vhd

```
-- Example 30: Fibonacci sequence
library IEEE;
use IEEE.STD_LOGIC_1164.all;
use IEEE.STD_LOGIC_unsigned.all;

entity fib is
       port(
              clk : in STD_LOGIC;
              clr : in STD_LOGIC;
              f : out STD_LOGIC_VECTOR(13 downto 0)
           );
end fib;

architecture fib of fib is
signal fn1, fn: STD_LOGIC_VECTOR(13 downto 0);
constant fib_max: integer := 9999;
begin

P1: process(clk, clr)
      begin
            if clr = '1' then
                  fn <= "00000000000000";
                  fn1 <= "00000000000001";
            elsif clk'event and clk = '1' then
                  if fn1 < fib_max then
                        fn1 <= fn1 + fn;
                        fn <= fn1;
                  else
                        fn <= "00000000000000";
                        fn1 <= "00000000000001";
                  end if;
            end if;
      end process;
      f <= fn;
end fib;
```

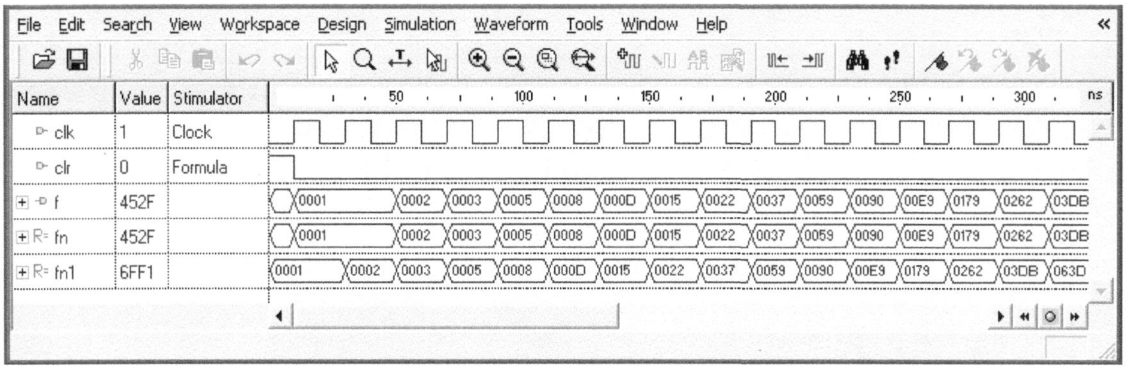

Figure 30.2 Simulation of the VHDL program in Listing 30.1

The adder is implemented with the + sign. Fig. 30.3 shows a top-level BDE design that can be implemented on the FPGA board. The Fibonacci sequence will increment at a 3 Hz rate. It will not take long to overflow the 4-digit hex display. Try it.

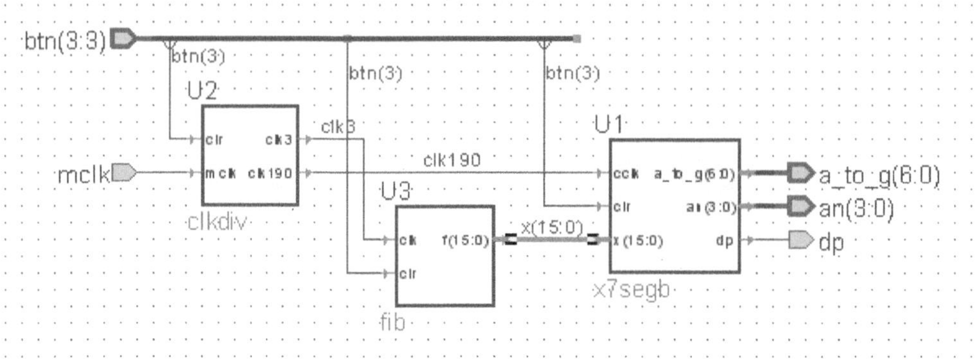

Figure 30.3 Top-level design for Fibonacci sequence

Appendix A

Aldec Active-HDL Tutorial

Part 1: Project Setup

Start the program by double-clicking the Active-HDL icon on the desktop.

Select *Create new workspace* and click *OK*.

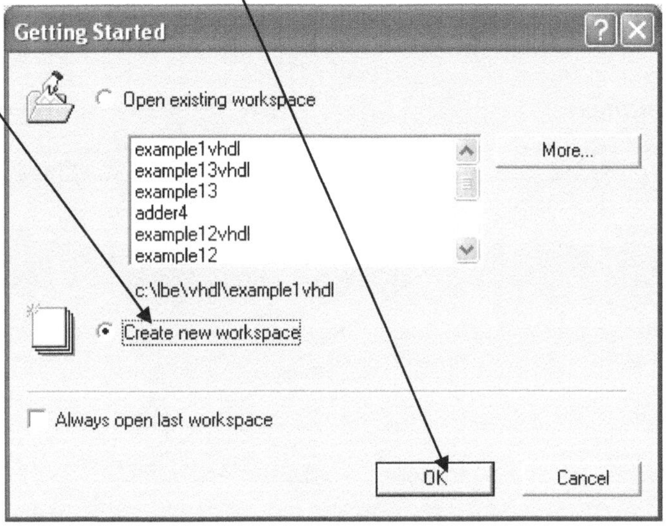

Browse to the directory where you want the project saved, type *Example1* for the workspace name and click *OK*.

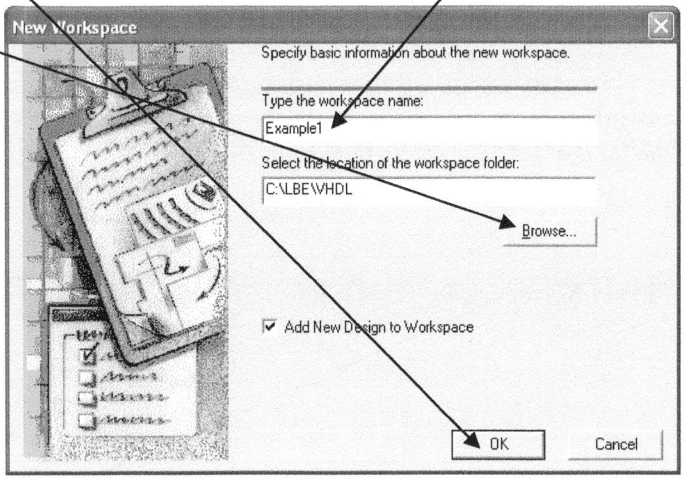

Select *Create an Empty Design with Design Flow* and click *Next.*

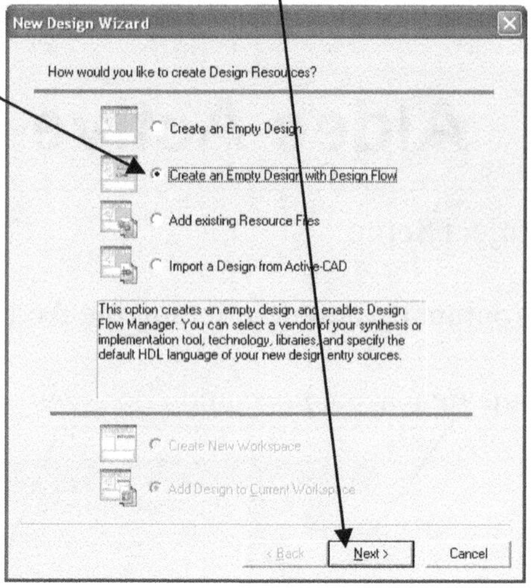

Click *Flow Settings*

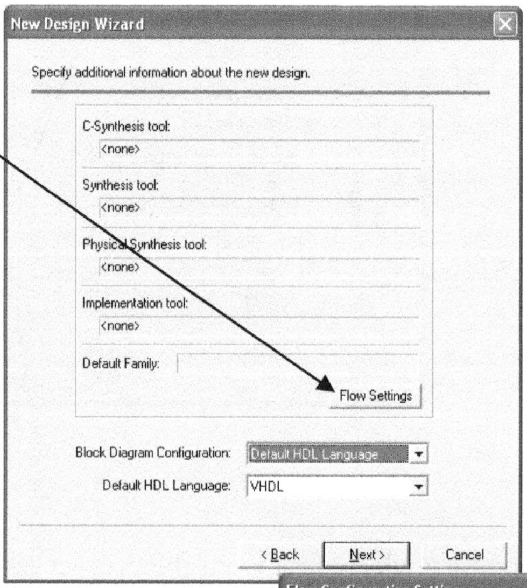

Select *HDL Synthesis*

Select *Xilinx ISE/WebPack 8.1 XST VHDL/Verilog*

Press *Select*

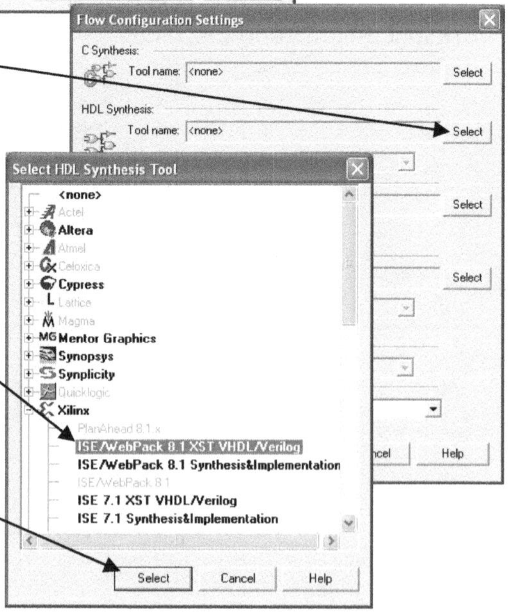

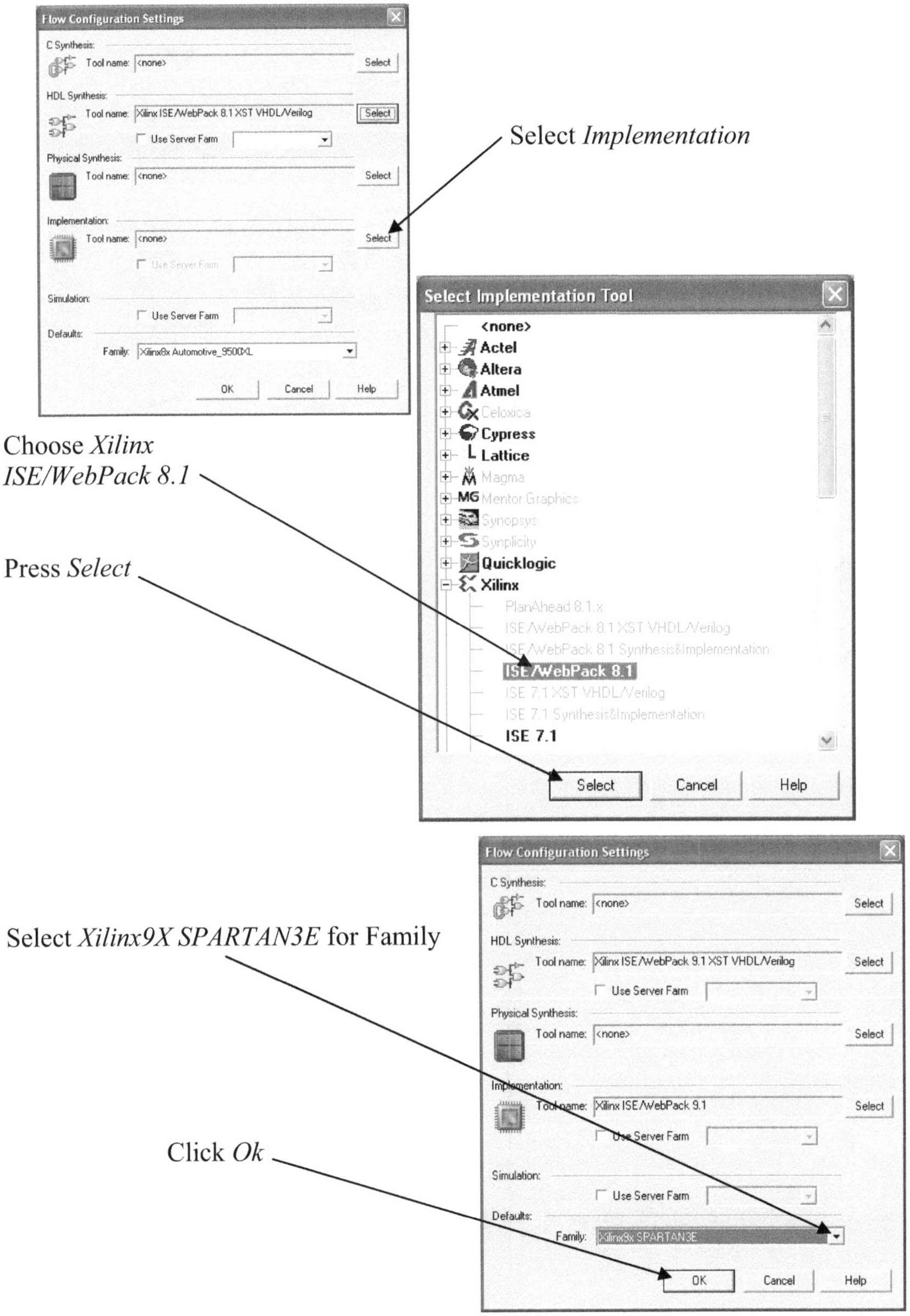
Flow Configuration Settings
Select *Implementation*
Select Implementation Tool
<none>
Actel
Altera
Atmel
Celoxica
Cypress
Lattice
Magma
Mentor Graphics
Synopsys
Synplicity
Quicklogic
Xilinx
PlanAhead 8.1.x
ISE/WebPack 8.1 XST VHDL/Verilog
ISE/WebPack 8.1 Synthesis&Implementation
ISE/WebPack 8.1
ISE 7.1 XST VHDL/Verilog
ISE 7.1 Synthesis&Implementation
ISE 7.1
Select
Cancel
Help
Choose *Xilinx ISE/WebPack 8.1*
Press *Select*
Select *Xilinx9X SPARTAN3E* for Family
Click *Ok*
Flow Configuration Settings
C Synthesis:
Tool name: <none>
HDL Synthesis:
Tool name: Xilinx ISE/WebPack 9.1 XST VHDL/Verilog
Use Server Farm
Physical Synthesis:
Tool name: <none>
Implementation:
Tool name: Xilinx ISE/WebPack 9.1
Use Server Farm
Simulation:
Use Server Farm
Defaults:
Family: Xilinx9x SPARTAN3E
OK
Cancel
Help

Select *VHDL* for the Default HDL Language

Click *Next*

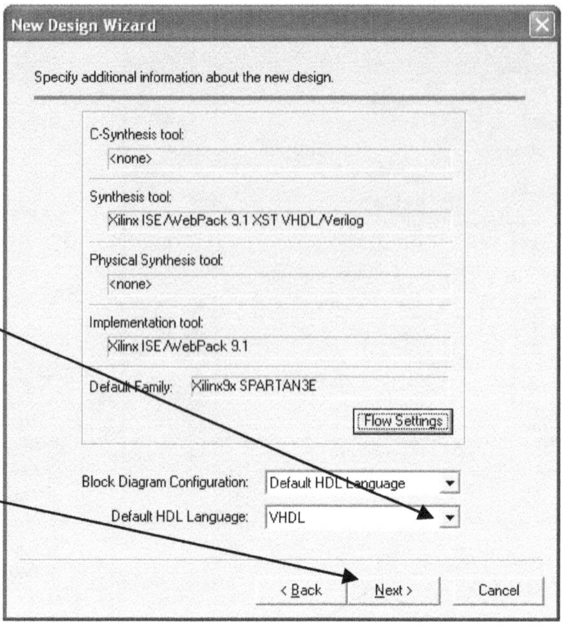

Type *swled* for the design name

and click *Next*.

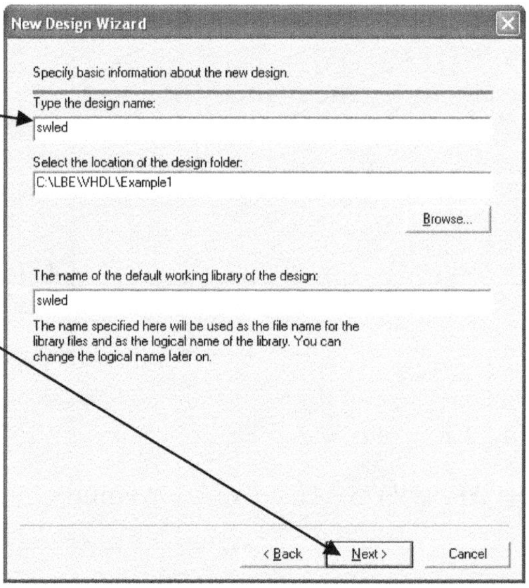

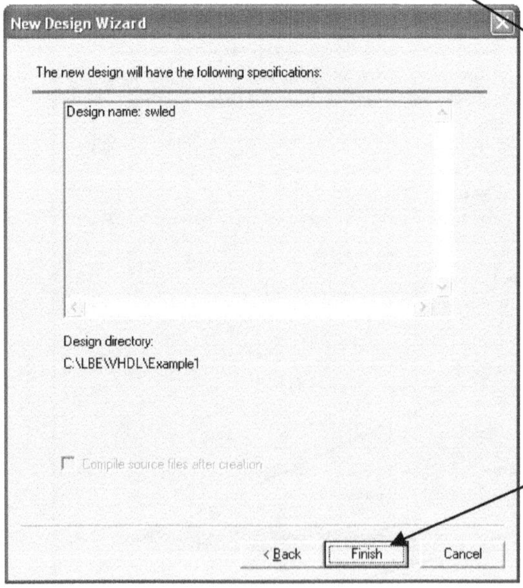

Click *Finish*.

Part 2: Design Entry – sw2led.bde

Click on *BDE*.

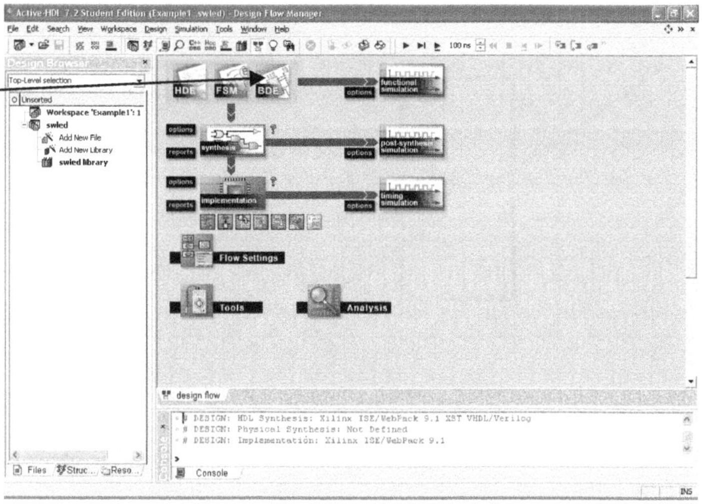

Click *Next*.

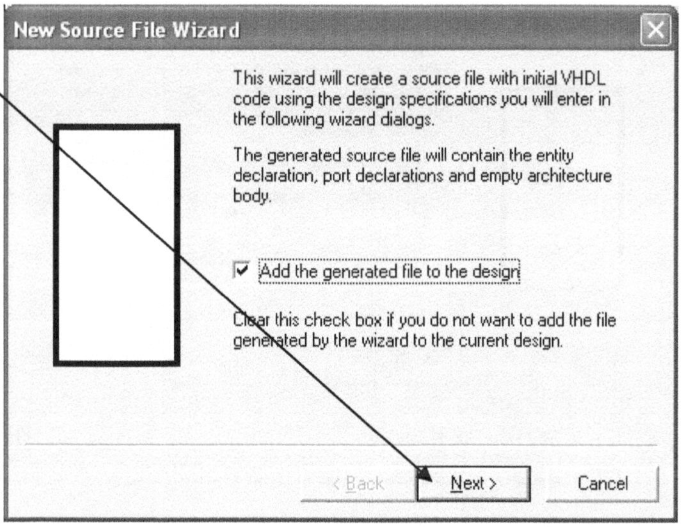

Select *VHDL* and Click *Next*

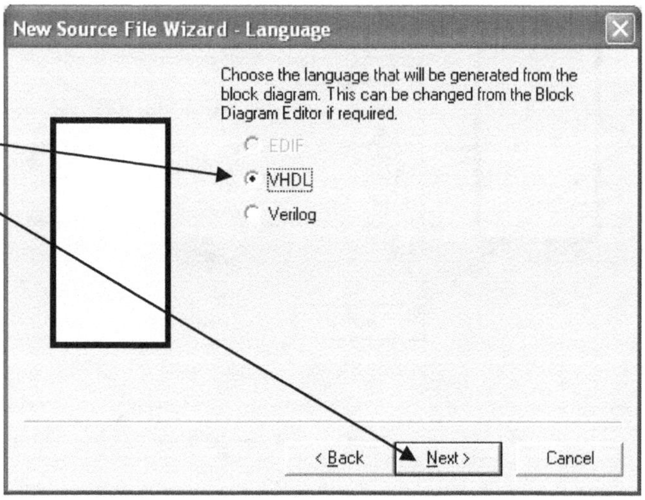

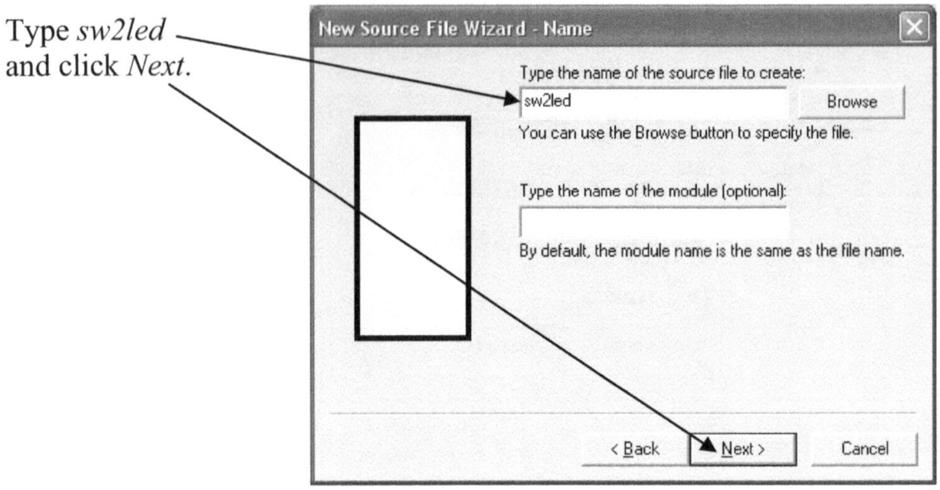
Type sw2led and click Next.
New Source File Wizard - Name
Type the name of the source file to create:
sw2led
Browse
You can use the Browse button to specify the file.
Type the name of the module (optional):
By default, the module name is the same as the file name.
< Back
Next >
Cancel

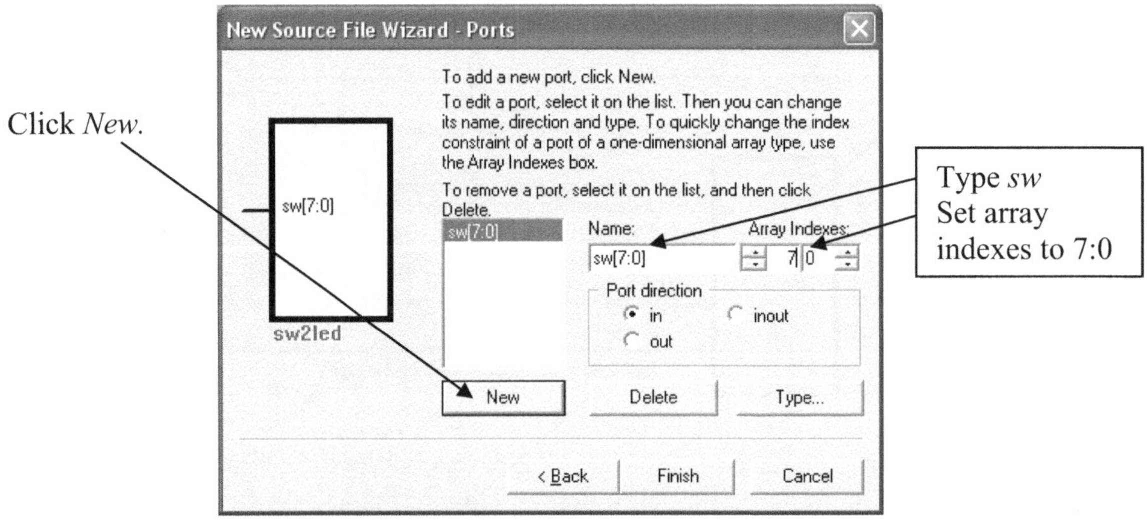
Click New.
New Source File Wizard - Ports
To add a new port, click New.
To edit a port, select it on the list. Then you can change its name, direction and type. To quickly change the index constraint of a port of a one-dimensional array type, use the Array Indexes box.
To remove a port, select it on the list, and then click Delete.
sw[7:0]
sw2led
Name:
Array Indexes:
sw[7:0]
7
0
Port direction
in
inout
out
New
Delete
Type...
< Back
Finish
Cancel
Type sw
Set array indexes to 7:0

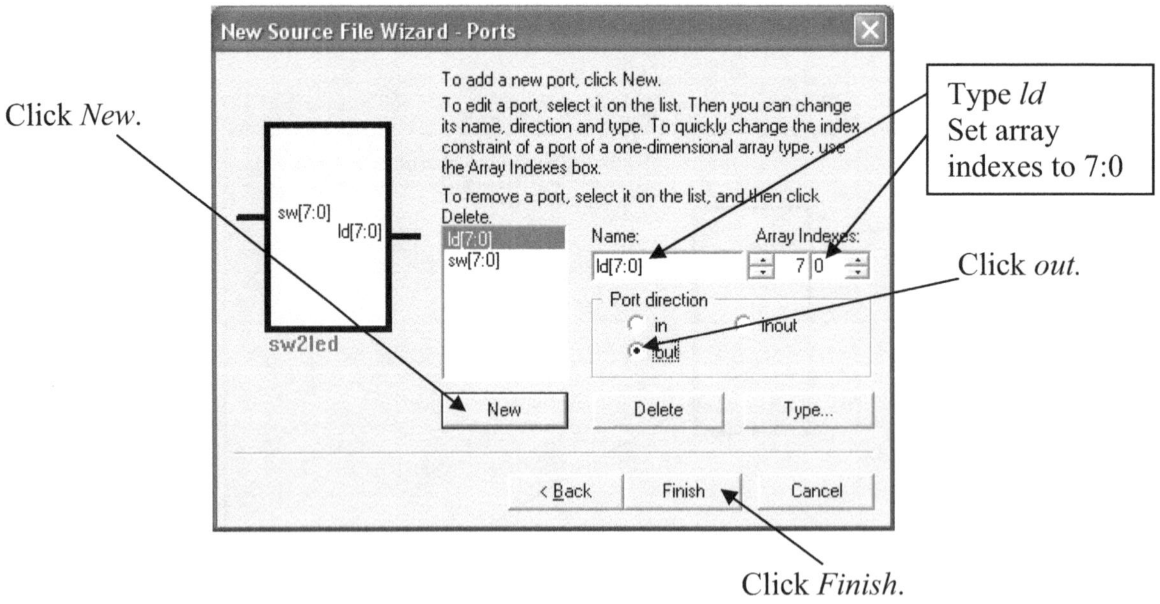
Click New.
New Source File Wizard - Ports
To add a new port, click New.
To edit a port, select it on the list. Then you can change its name, direction and type. To quickly change the index constraint of a port of a one-dimensional array type, use the Array Indexes box.
To remove a port, select it on the list, and then click Delete.
sw[7:0]
ld[7:0]
sw2led
ld[7:0]
sw[7:0]
Name:
Array Indexes:
ld[7:0]
7
0
Port direction
in
inout
out
New
Delete
Type...
< Back
Finish
Cancel
Type ld
Set array indexes to 7:0
Click out.
Click Finish.

This will generate a block diagram (schematic) template with the *input* and *output* ports displayed.

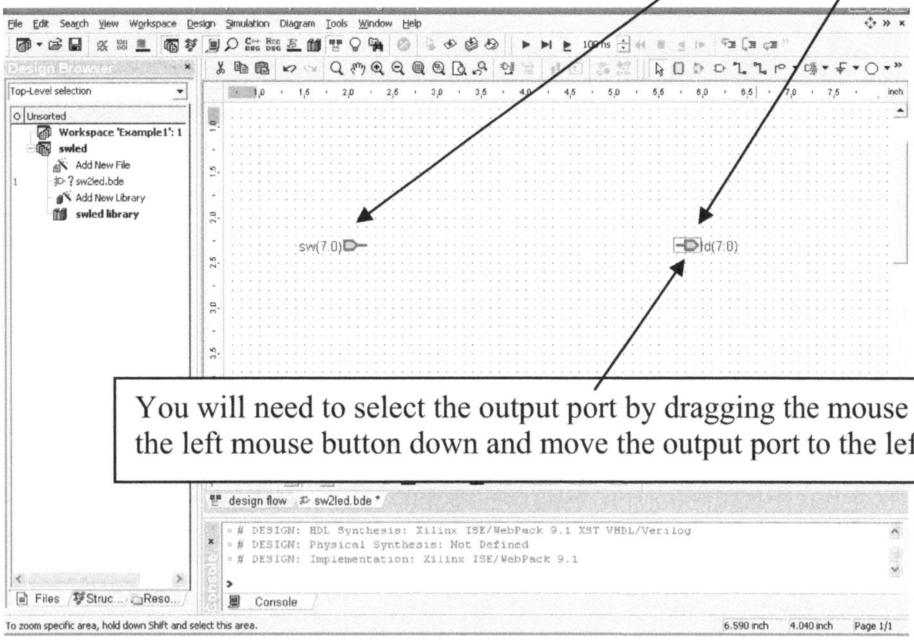

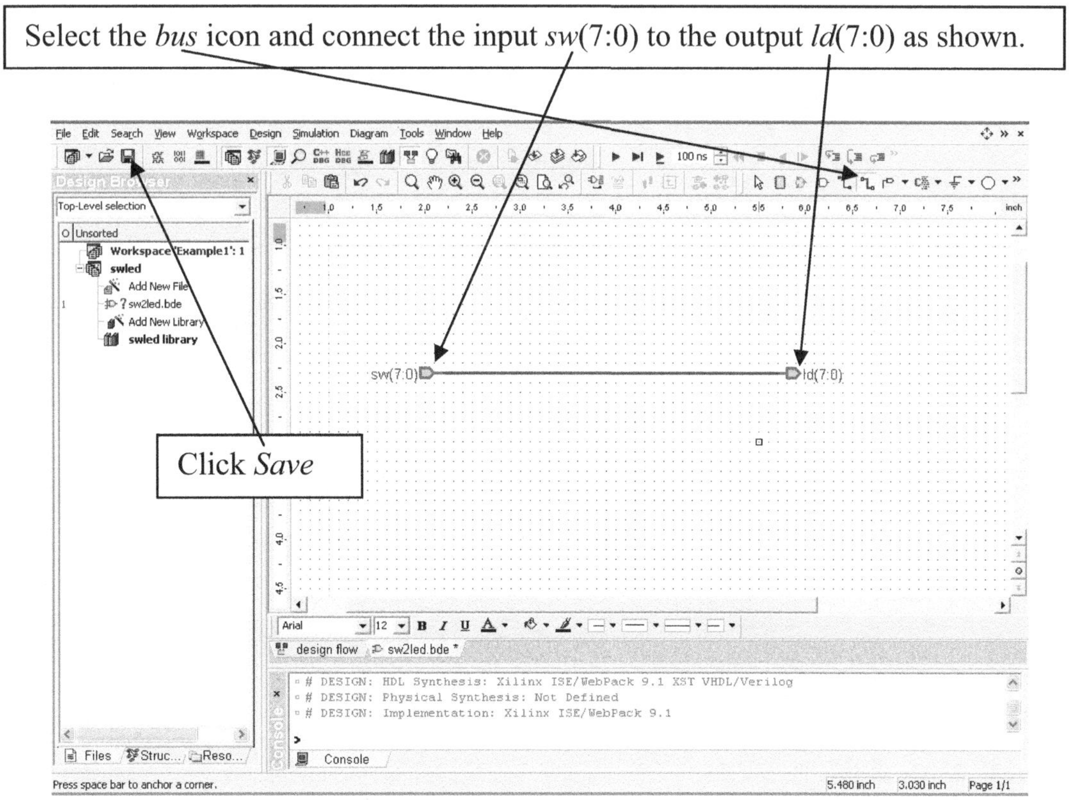

Right-click on *sw2led.bde* and select *Compile*

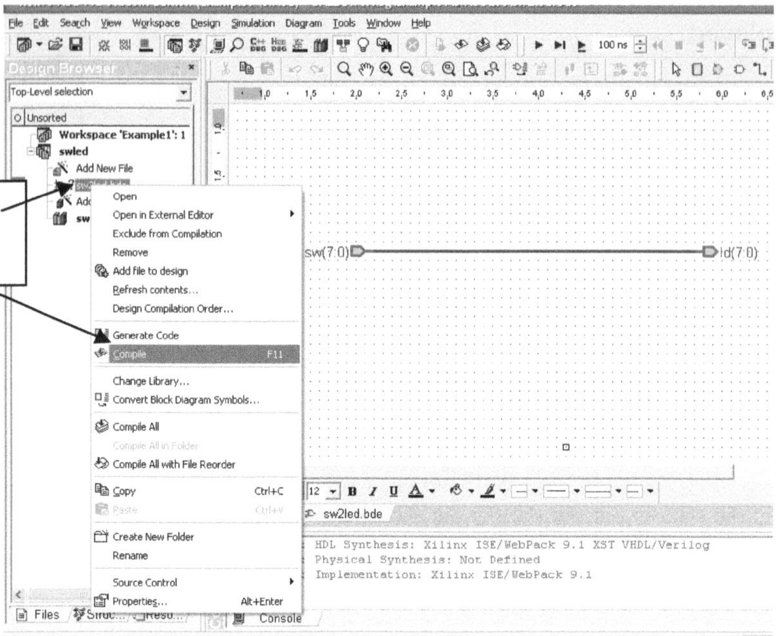

Part 3: Synthesis and Implementation

Click *design flow*

Click synthesis *options*

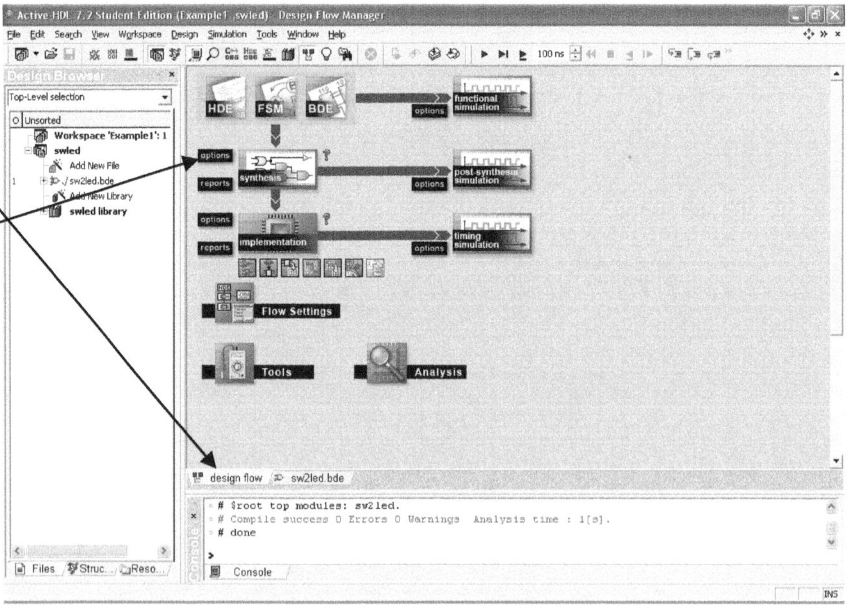

Pull down menu and select *sw2led* for Top-level Unit.

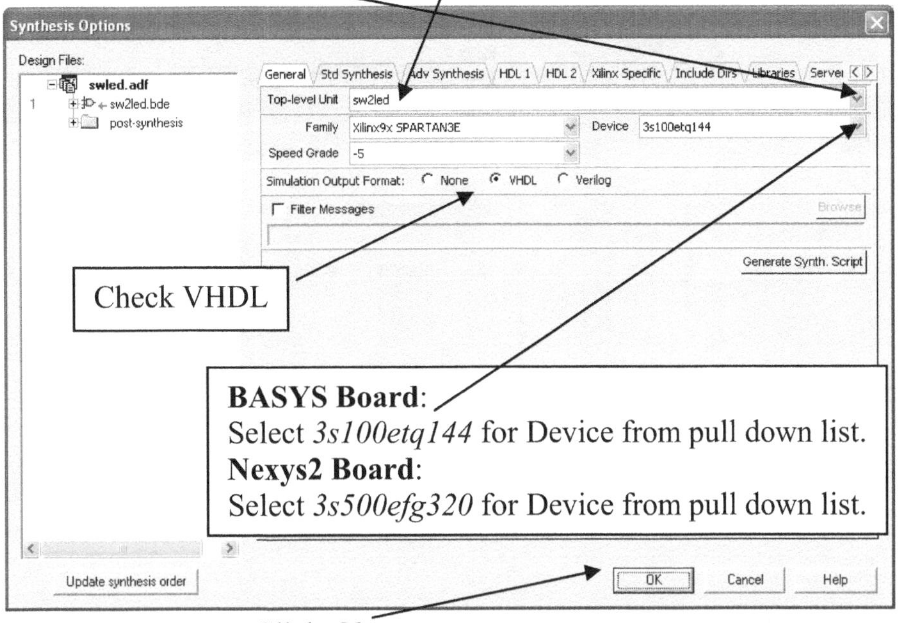

Click *Ok*.

Click *synthesis*

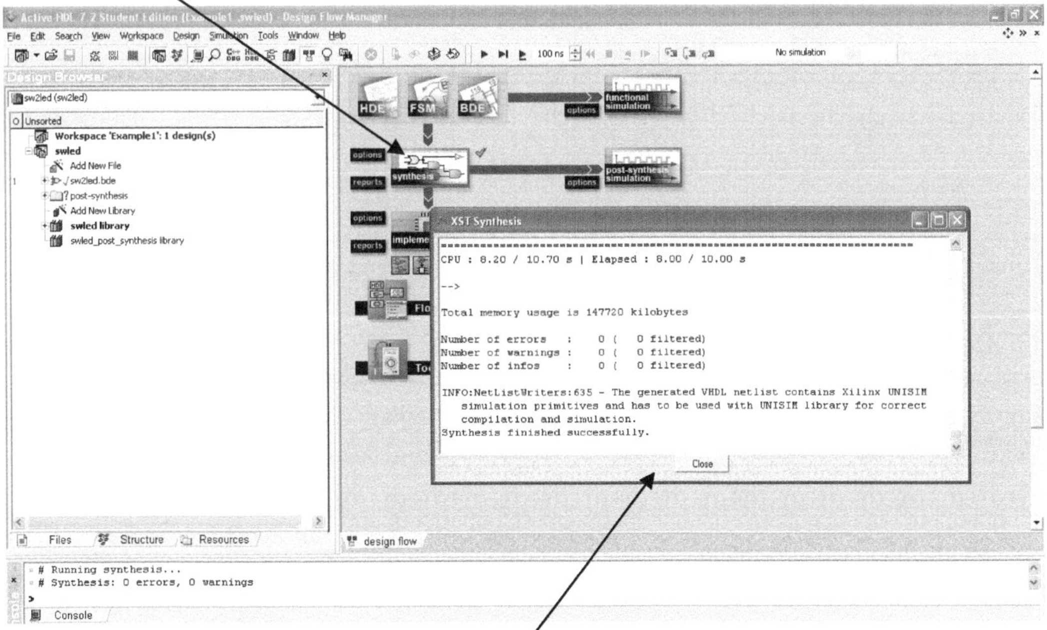

After synthesis is complete, click *Close*.

Click *implementation options*

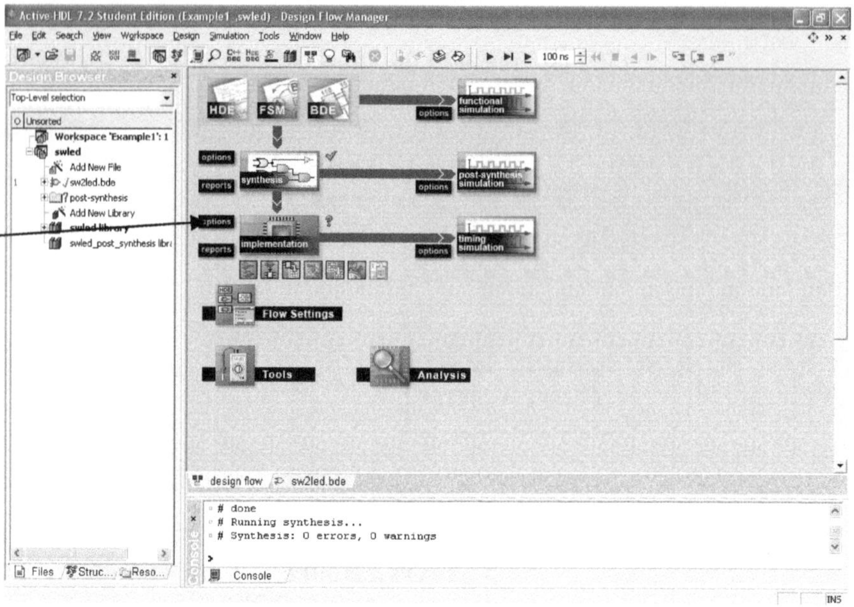

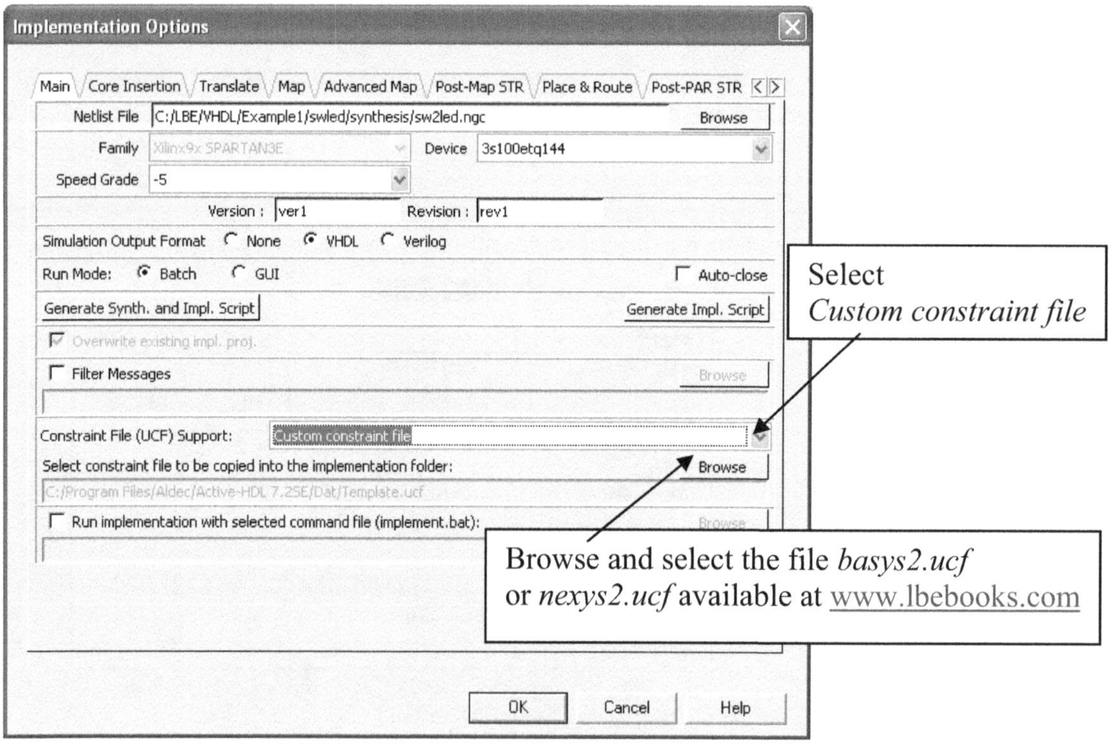

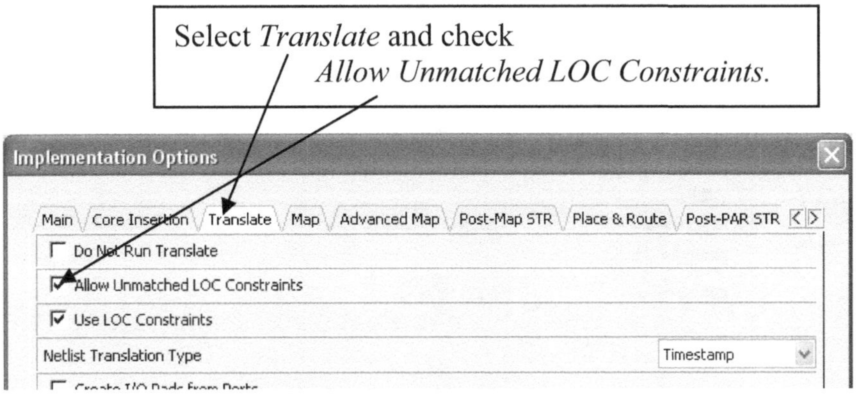
Select *Translate* and check *Allow Unmatched LOC Constraints.*
Implementation Options
Main
Core Insertion
Translate
Map
Advanced Map
Post-Map STR
Place & Route
Post-PAR STR
Do Not Run Translate
Allow Unmatched LOC Constraints
Use LOC Constraints
Netlist Translation Type
Timestamp

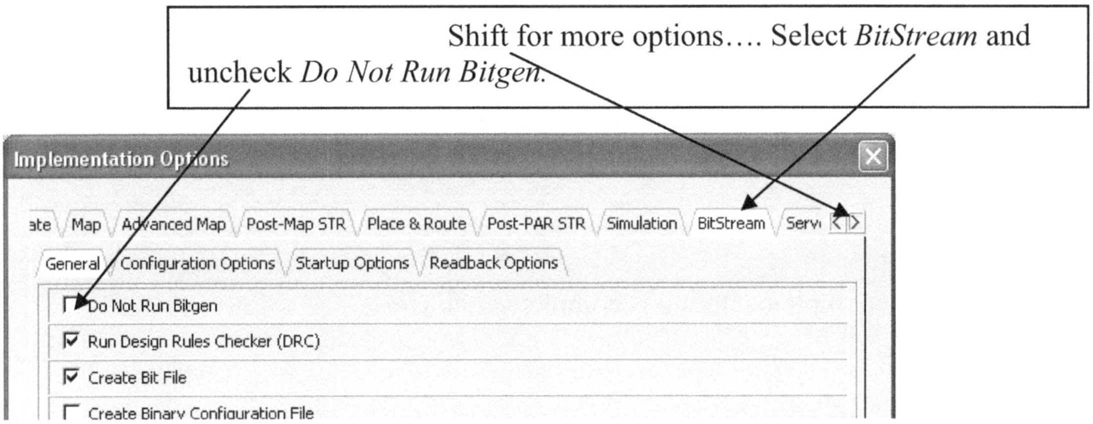
Shift for more options…. Select *BitStream* and uncheck *Do Not Run Bitgen.*
Implementation Options
Map
Advanced Map
Post-Map STR
Place & Route
Post-PAR STR
Simulation
BitStream
General
Configuration Options
Startup Options
Readback Options
Do Not Run Bitgen
Run Design Rules Checker (DRC)
Create Bit File
Create Binary Configuration File

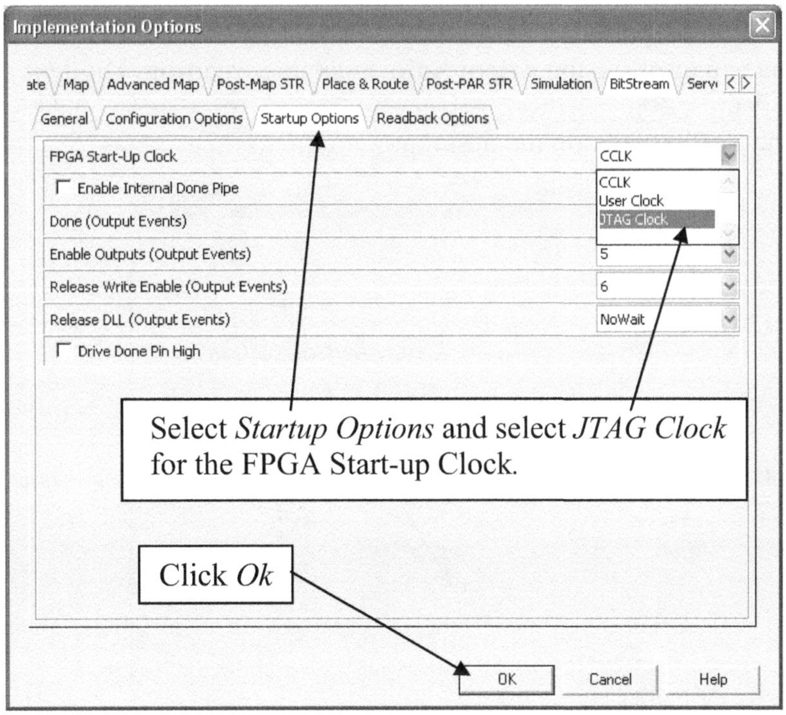
Implementation Options
Map
Advanced Map
Post-Map STR
Place & Route
Post-PAR STR
Simulation
BitStream
General
Configuration Options
Startup Options
Readback Options
FPGA Start-Up Clock
CCLK
Enable Internal Done Pipe
CCLK
User Clock
JTAG Clock
Done (Output Events)
Enable Outputs (Output Events)
5
Release Write Enable (Output Events)
6
Release DLL (Output Events)
NoWait
Drive Done Pin High
Select *Startup Options* and select *JTAG Clock* for the FPGA Start-up Clock.
Click *Ok*
OK
Cancel
Help

Click *implementation*

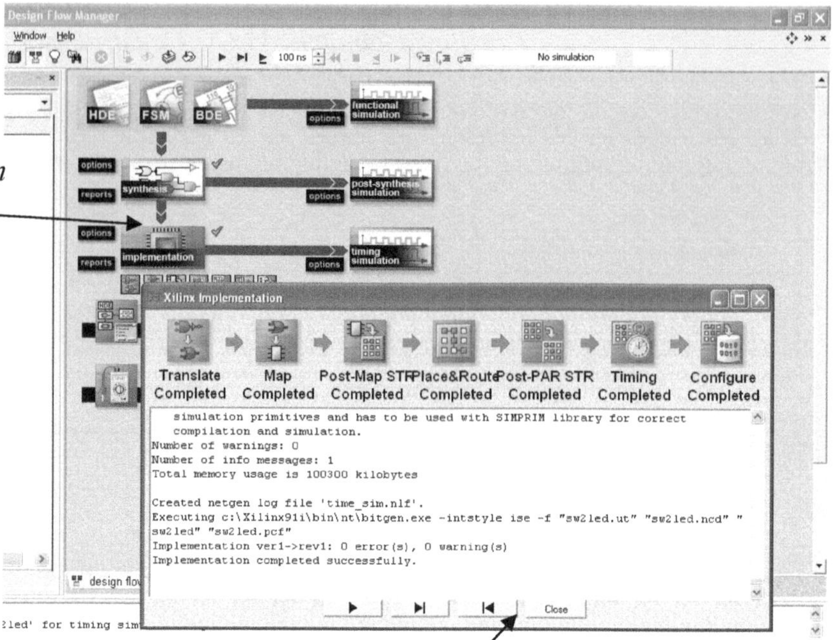

When implementation is complete click *Close.*

Part 4: Program FPGA Board

To program the Spartan3E on the BASYS or Nexys-2 boards we will use the **ExPort** tool that is part of the the **Adept Suite** available free from Digilent at http://www.digilentinc.com/Software/Adept.cfm?Nav1=Software&Nav2=Adept Double-click the **ExPort** icon on the desktop.

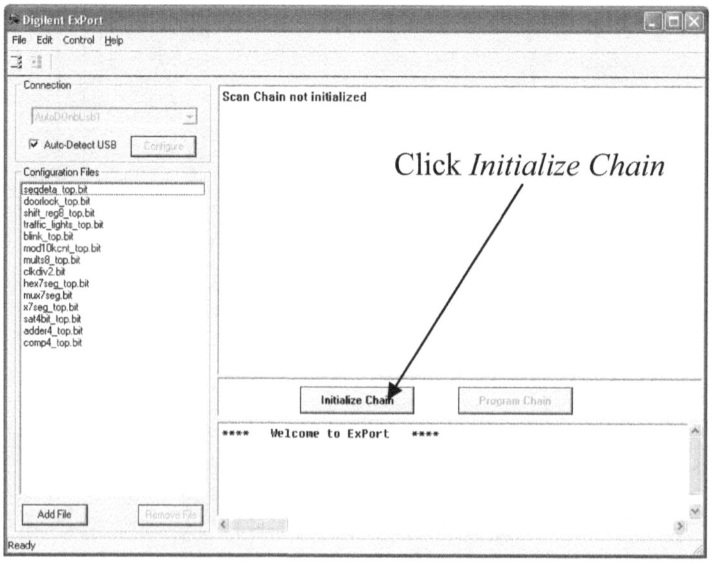

Click *Browse* and go to *Example1->swled->implement->ver1->rev1->sw2led.bit*

Select *sw2led.bit*

Click *Program Chain*

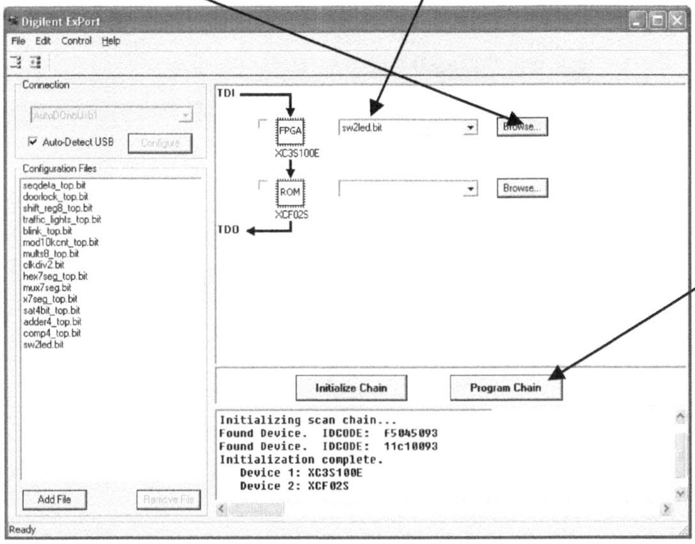

Your program is now running on the board. Change the switches and watch the LEDs.

Part 5: Design Entry – gates2.bde

Click on *BDE*.

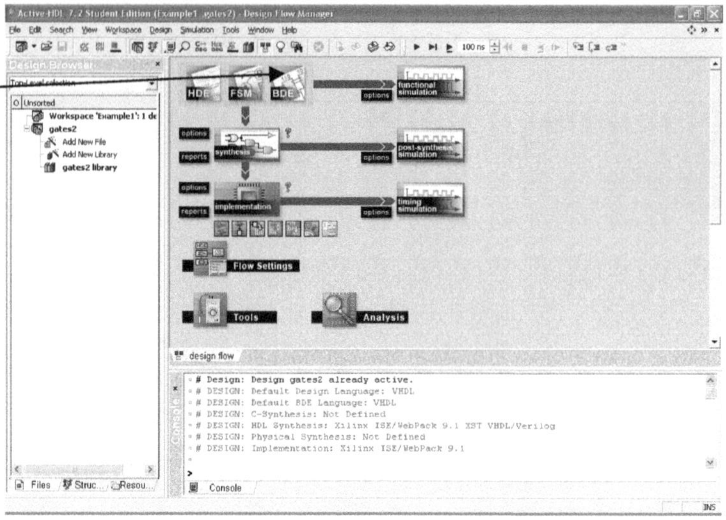

Click *Next*.

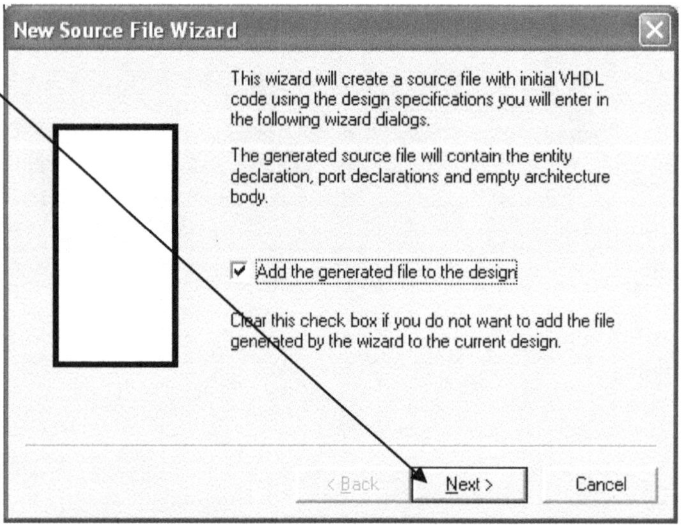

Select *VHDL* and Click *Next*

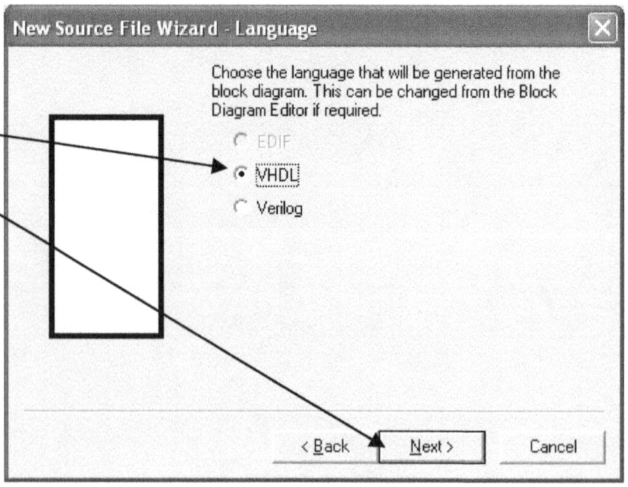

Type *gates2* and click *Next.*

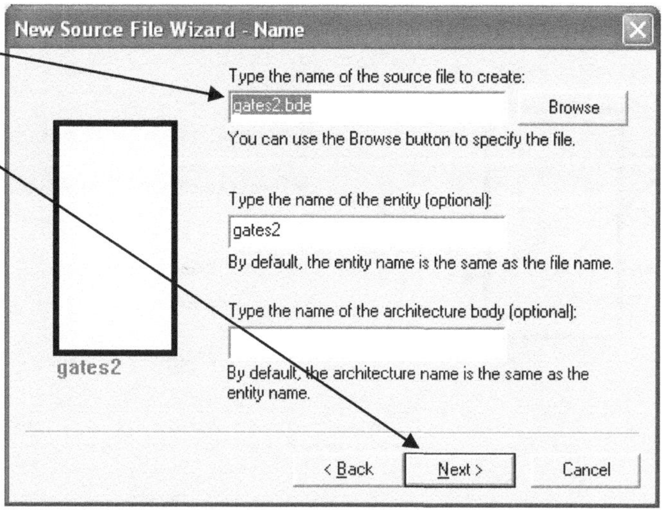

Click *New.*

Type *a.*

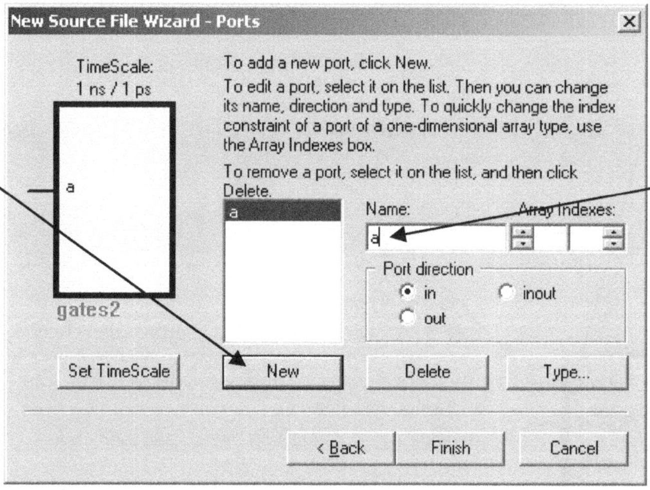

Click *New.*

Type *b.*

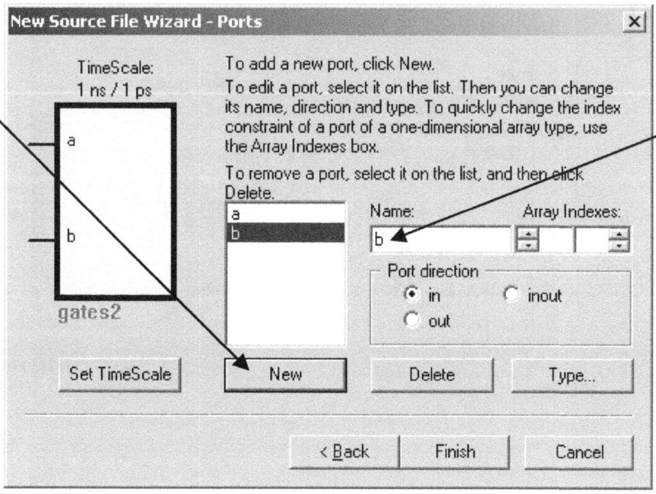

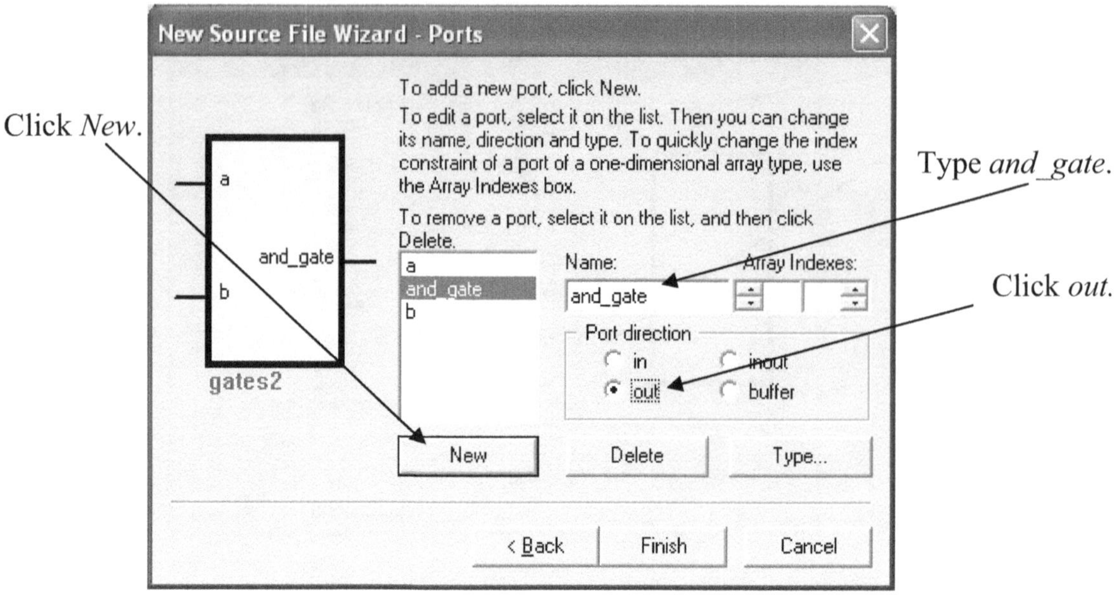

Continue to click *New* and add the outputs *nand_gate*, *or_gate*, *nor_gate*, *xor_gate*, and *xnor_gate*.

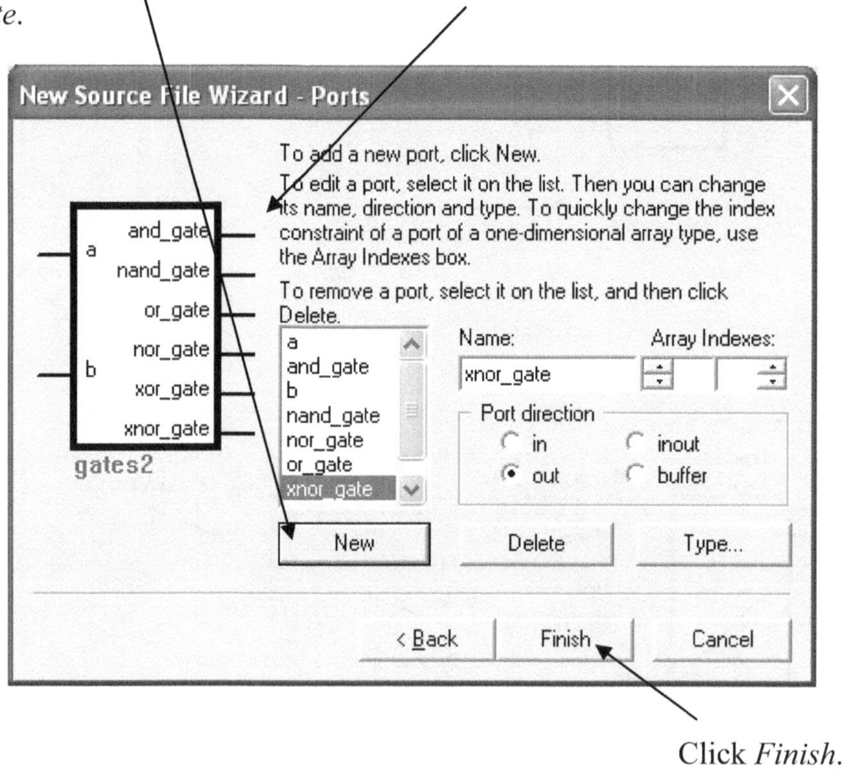

This will generate a block diagram (schematic) template with the *input* and *output* ports displayed.

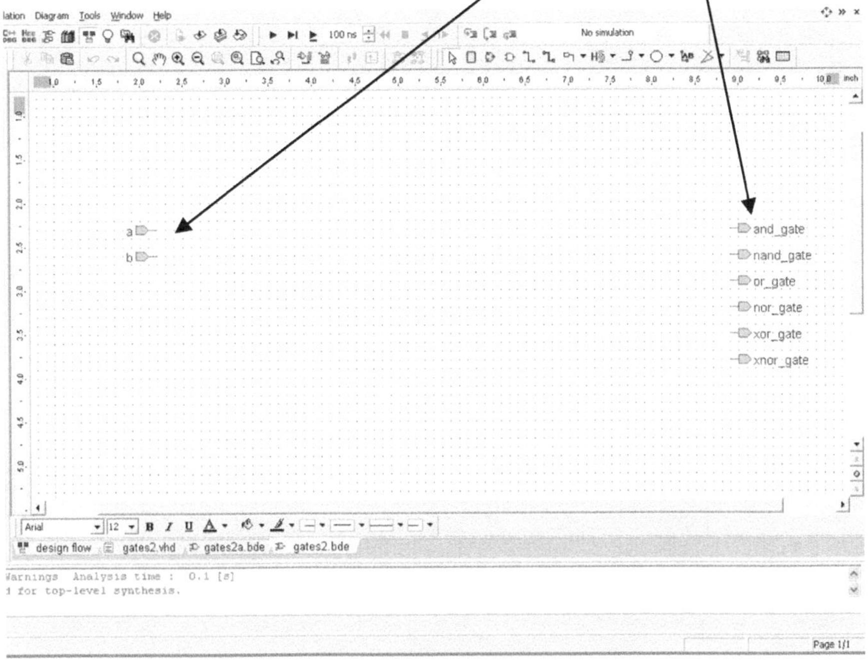

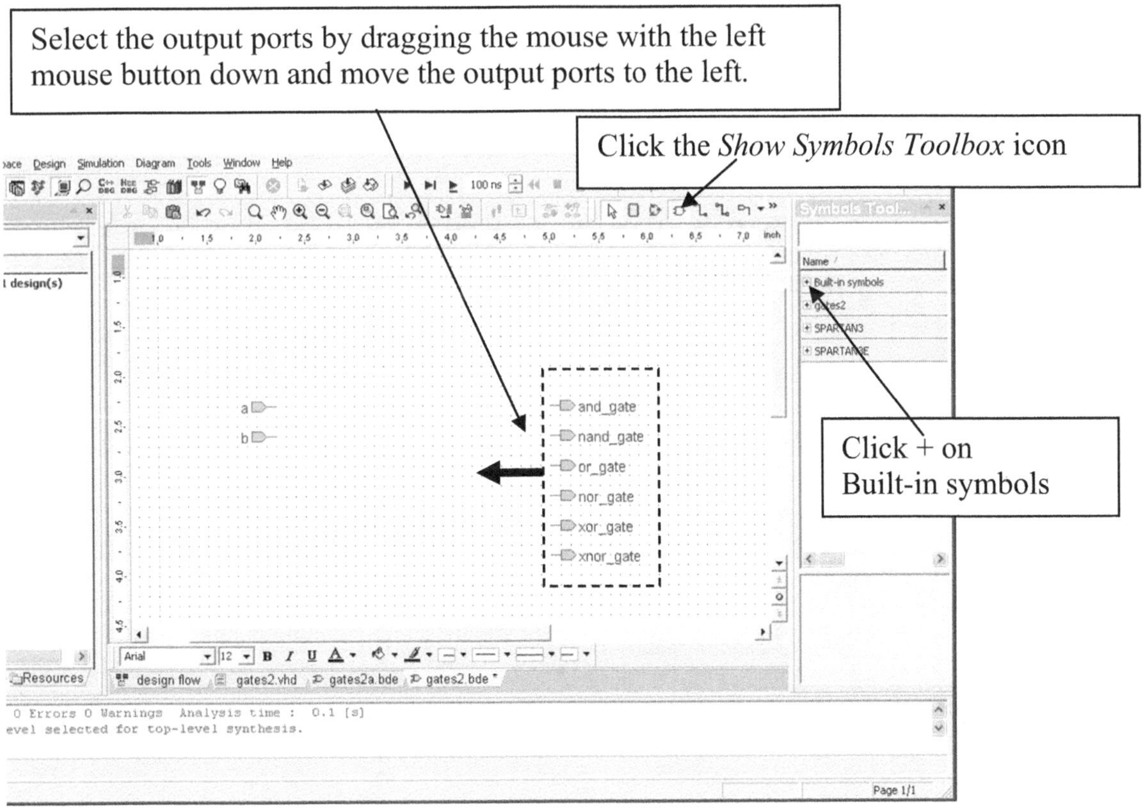

Grab the *and2* symbol with the mouse and drag it to the output port *and_gate*

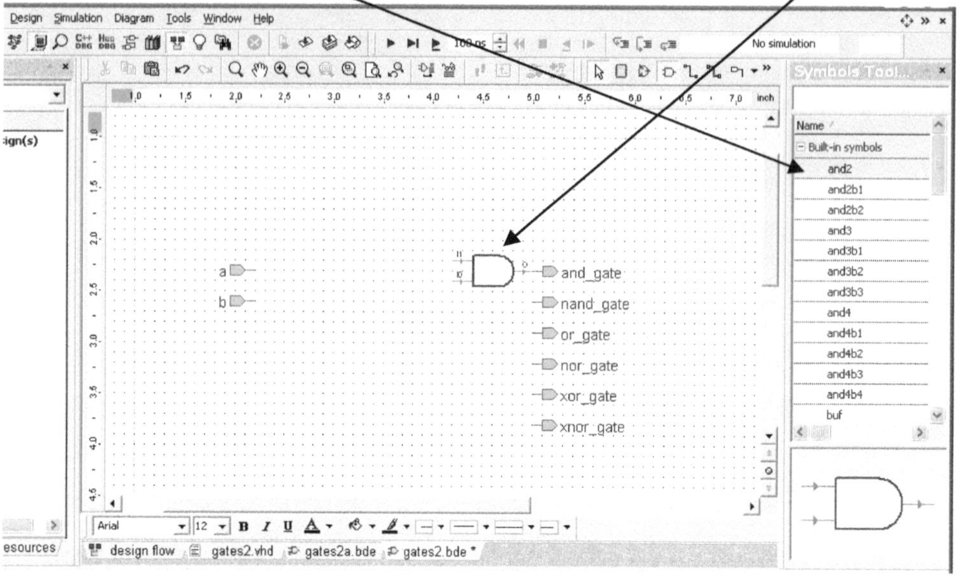

Grab the symbols for *nand2*, *or2*, *nor2*, *xor2*, and *xnor2* and drag them to the appropriate output port, moving the output ports down as necessary.

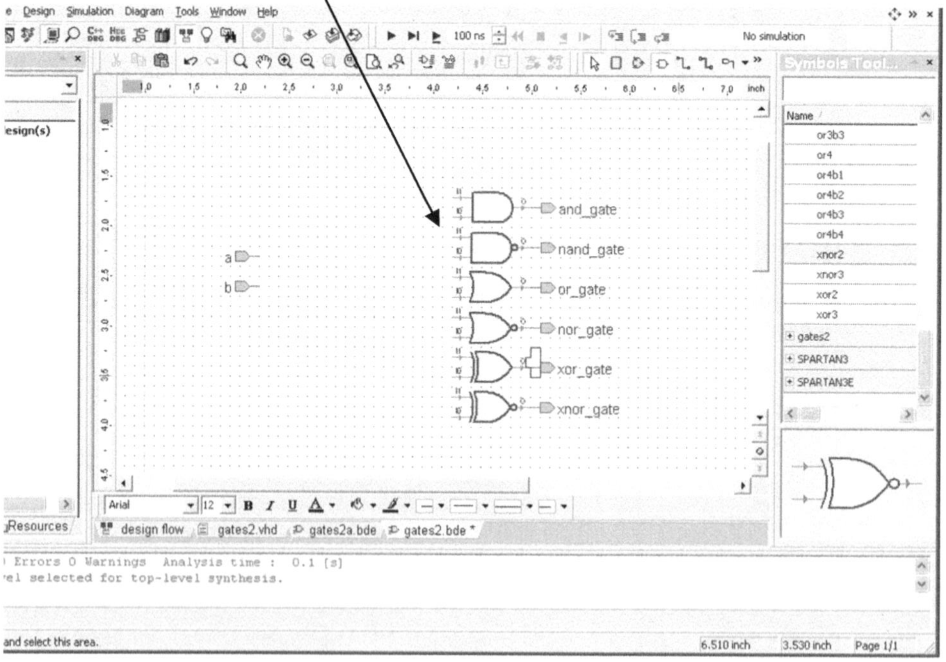

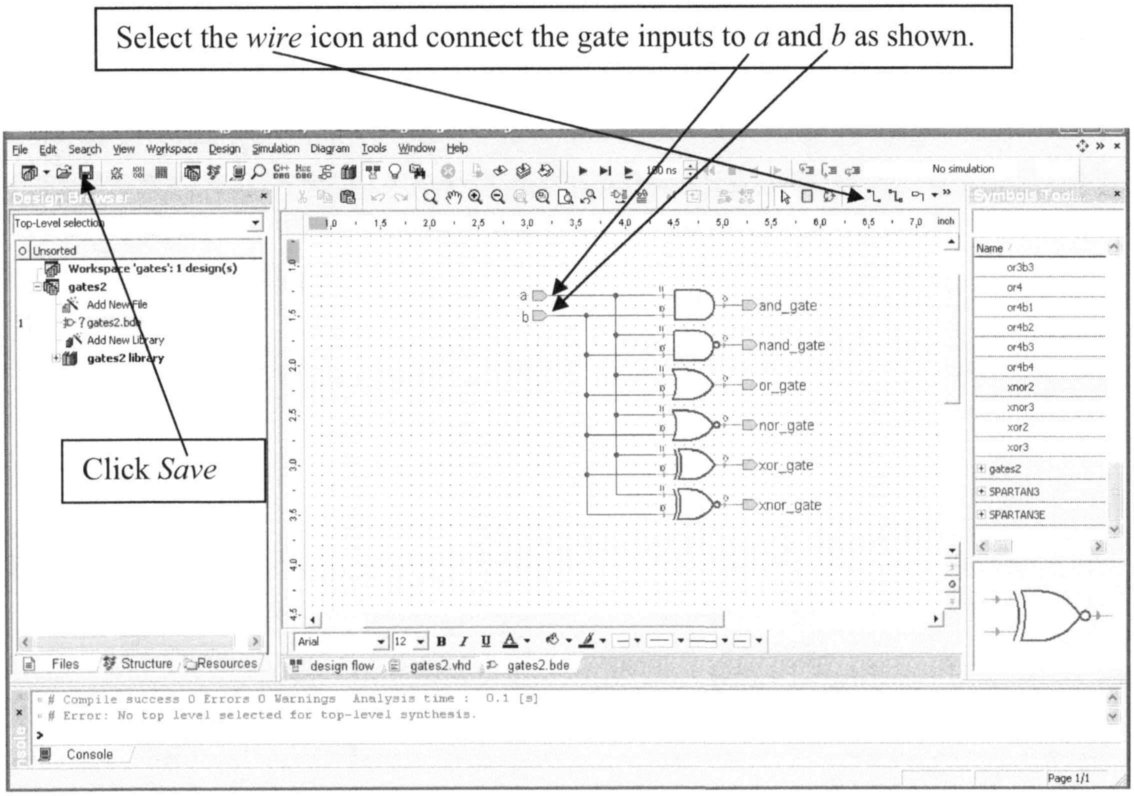
Select the *wire* icon and connect the gate inputs to *a* and *b* as shown.
Click *Save*
File Edit Search View Workspace Design Simulation Diagram Tools Window Help
No simulation
Design Browser
Top-Level selection
Workspace 'gates': 1 design(s)
gates2
Add New File
gates2.bde
Add New Library
gates2 library
a
b
and_gate
nand_gate
or_gate
nor_gate
xor_gate
xnor_gate
Name
or3b3
or4
or4b1
or4b2
or4b3
or4b4
xnor2
xnor3
xor2
xor3
gates2
SPARTAN3
SPARTAN3E
Files
Structure
Resources
design flow
gates2.vhd
gates2.bde
Compile success 0 Errors 0 Warnings Analysis time : 0.1 [s]
Error: No top level selected for top-level synthesis.
Console
Page 1/1

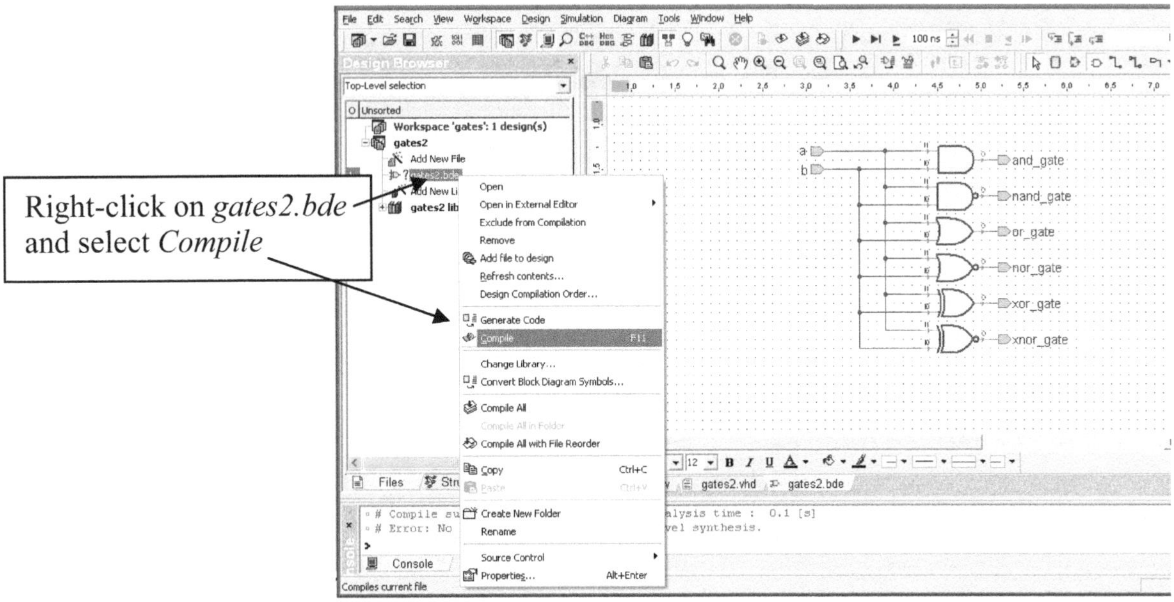
Right-click on *gates2.bde* and select *Compile*
Open
Open in External Editor
Exclude from Compilation
Remove
Add file to design
Refresh contents...
Design Compilation Order...
Generate Code
Compile F11
Change Library...
Convert Block Diagram Symbols...
Compile All
Compile All in Folder
Compile All with File Reorder
Copy Ctrl+C
Paste Ctrl+V
Create New Folder
Rename
Source Control
Properties... Alt+Enter
and_gate
nand_gate
or_gate
nor_gate
xor_gate
xnor_gate
Compiles current file

Part 6: Simulation

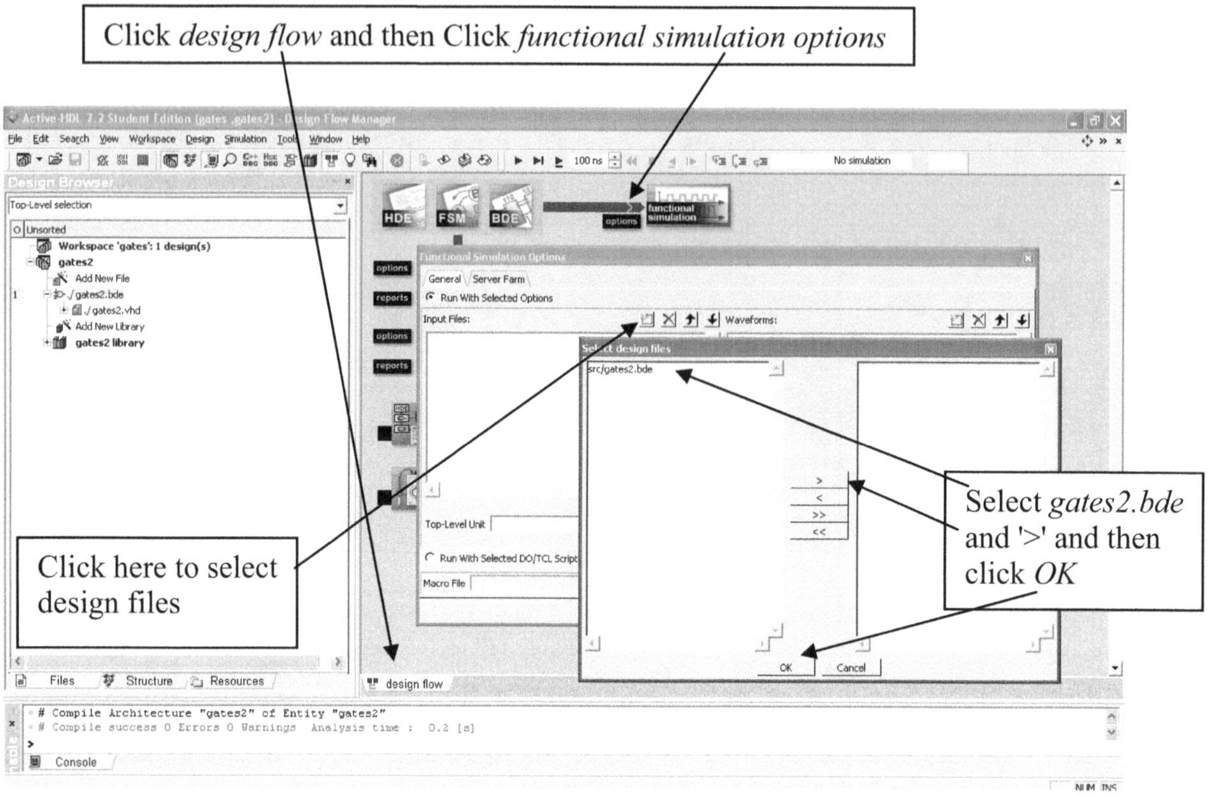

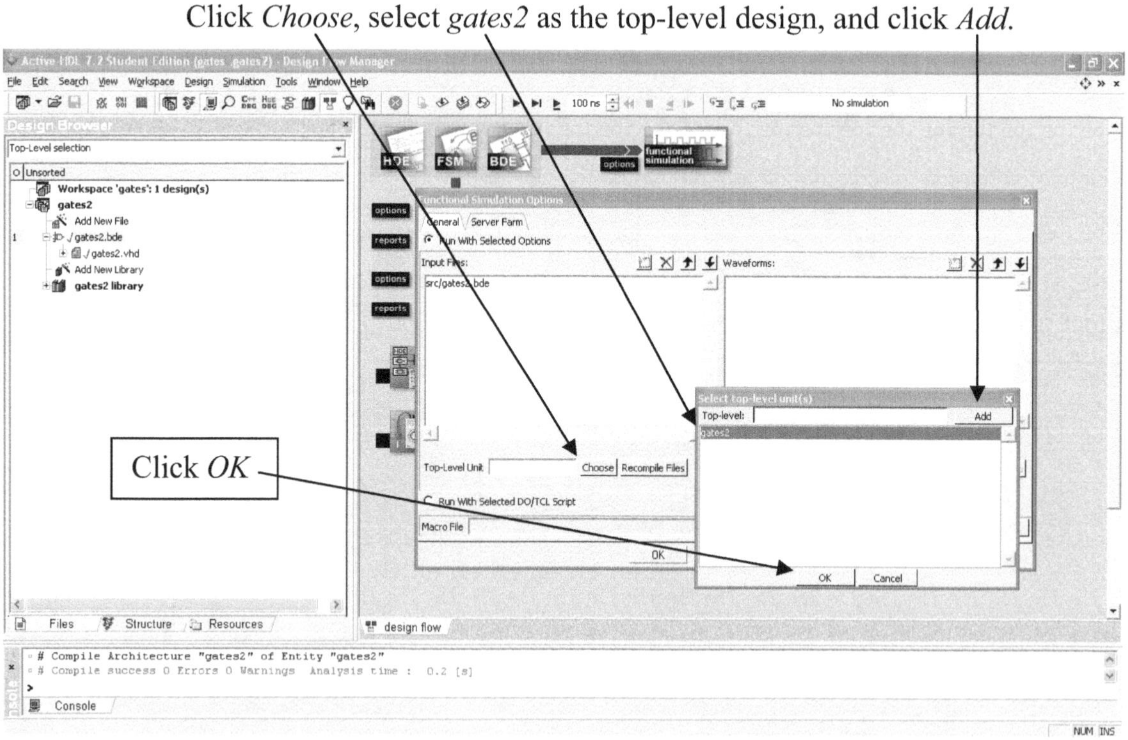

Click *Use Default Waveform*

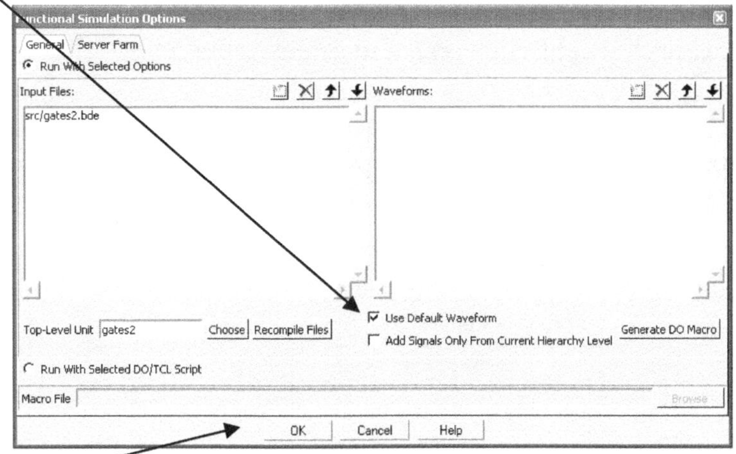

Click *OK*

Click *functional simulation*

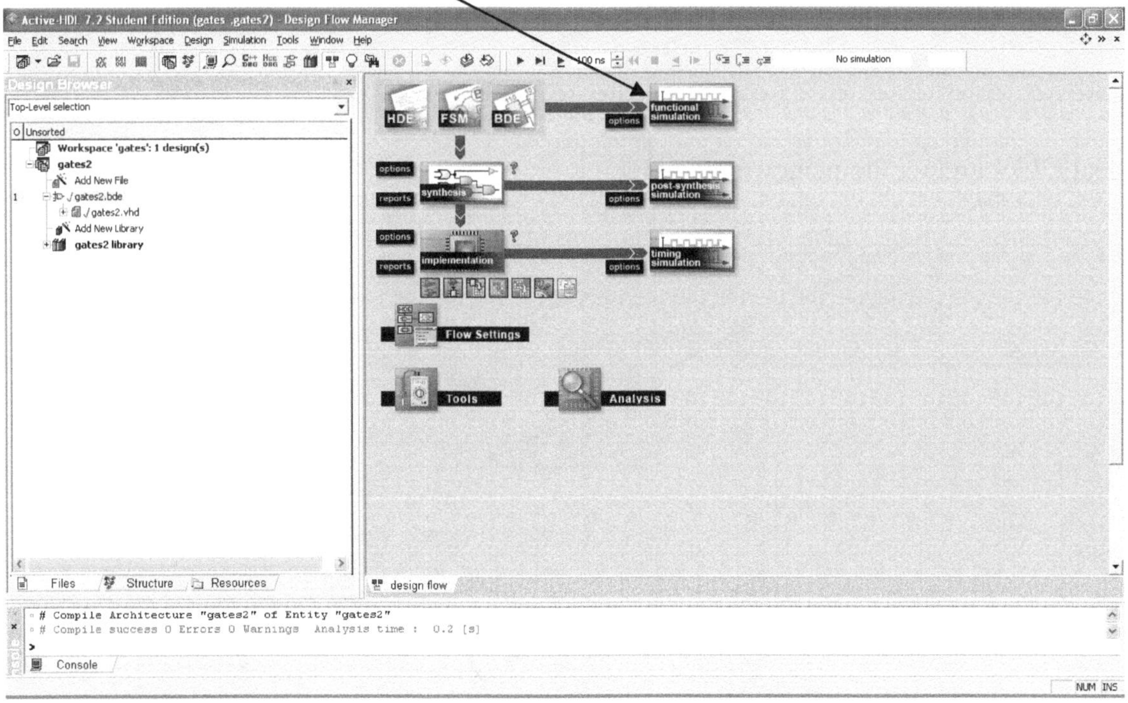

The waveform window will automatically come up with the simulation already initialized. Make sure the order is *a*, *b*, *and_*, *nand_*, *or_*, *nor_*, *xor_*, *xnor* (grab and drag if necessary). Right-click on *a* and select *Stimulators*.

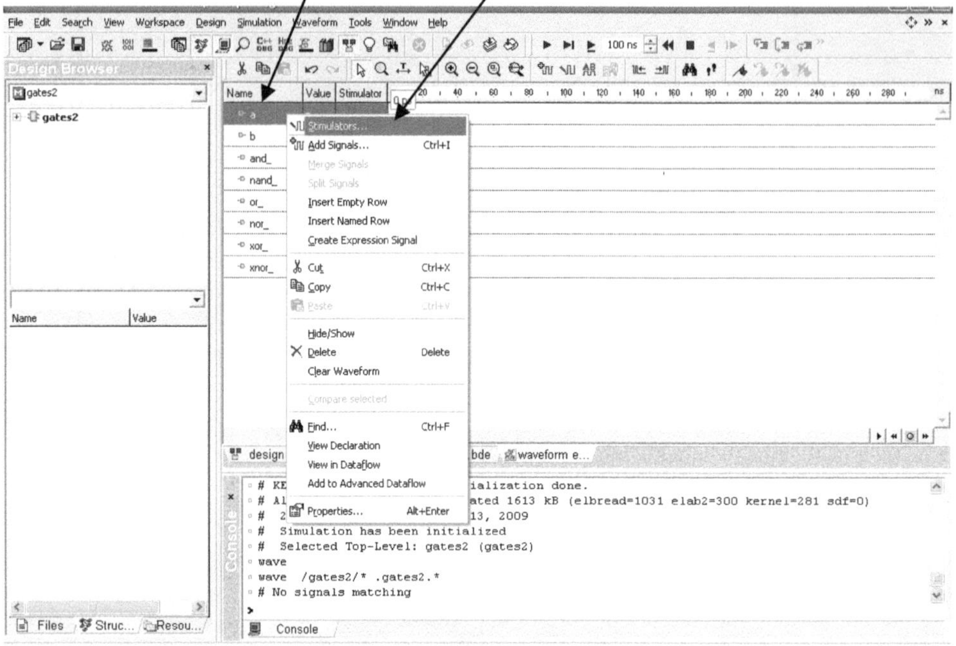

Select *Clock* and set Frequency to 25 MHz

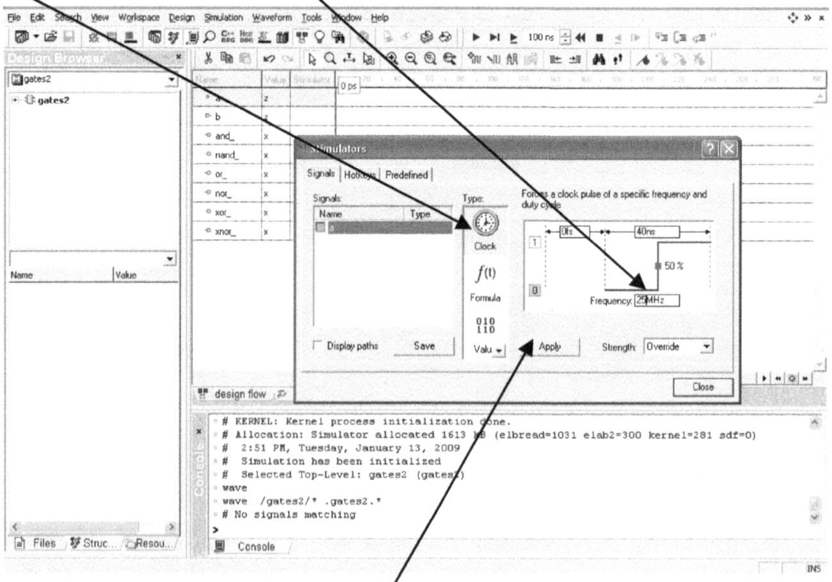

Click *Apply*

Click on *b*, select *Clock* and set Frequency to 50 MHz

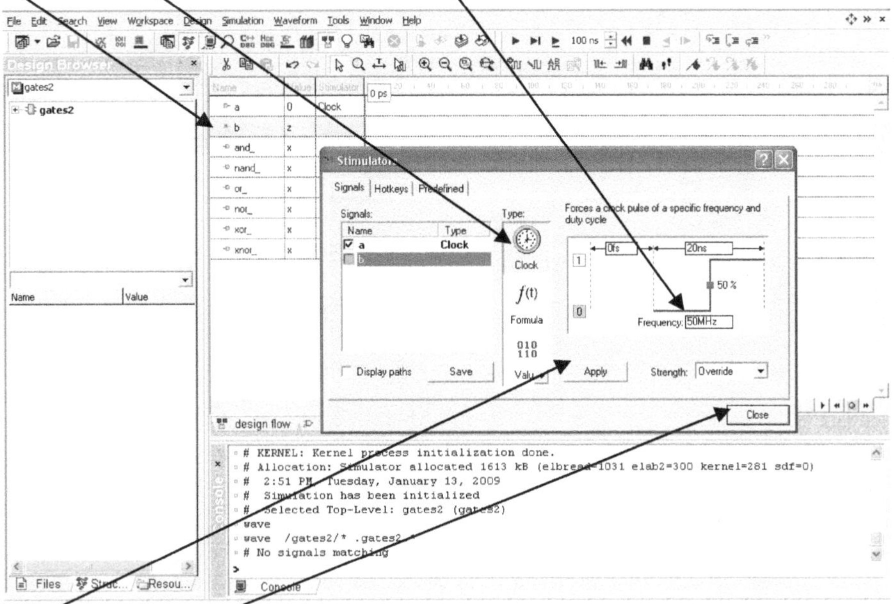

Click *Apply*

Click *Close*

Set simulation time to 200 ns

Click here to run simulation

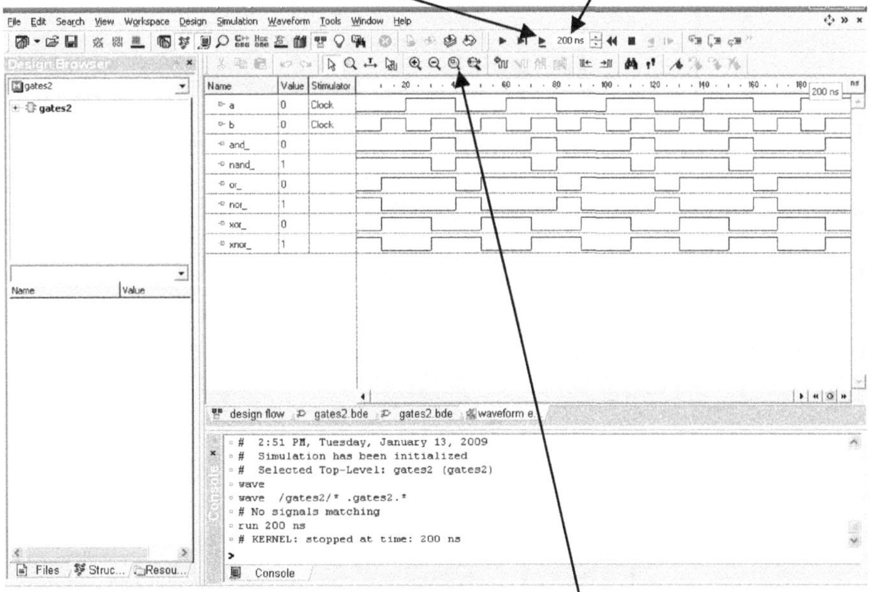

Click *Zoom to Fit.*

Part 7: Design Entry - HDE

Click on HDE.

Select *VHDL* and Click *OK*.

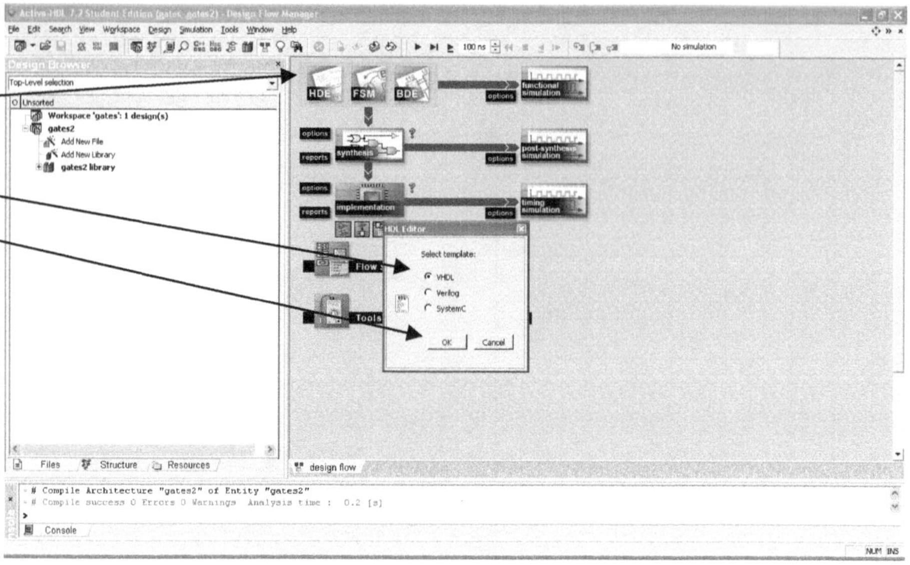

Click Next.

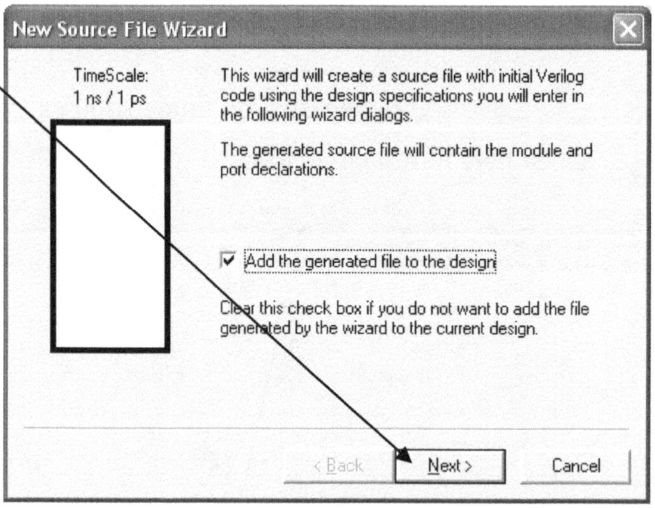

Type *gates2* and click *Next*.

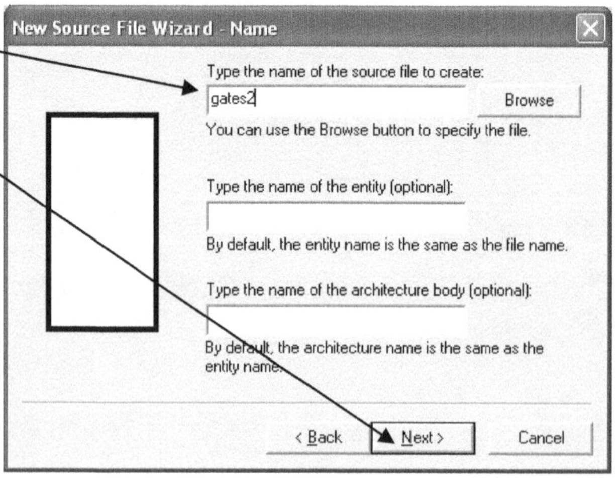

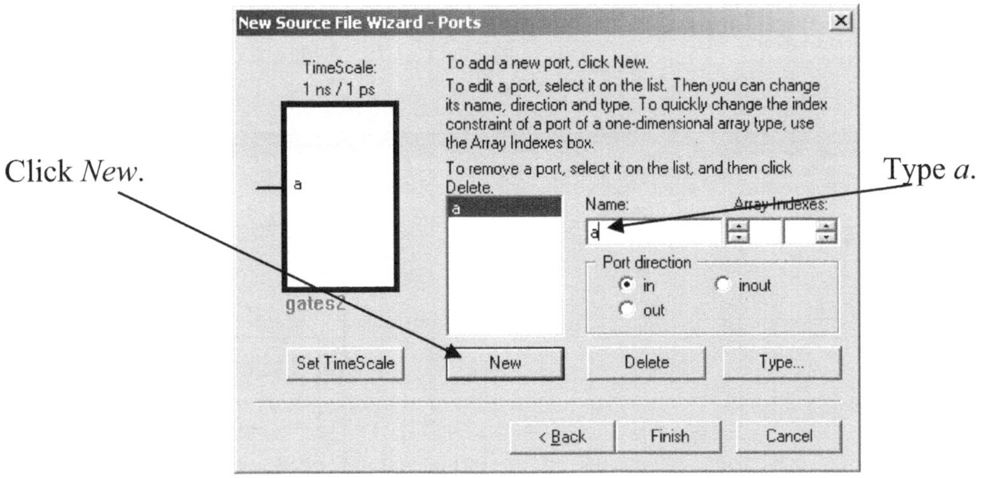
New Source File Wizard - Ports
TimeScale:
1 ns / 1 ps
To add a new port, click New.
To edit a port, select it on the list. Then you can change its name, direction and type. To quickly change the index constraint of a port of a one-dimensional array type, use the Array Indexes box.
To remove a port, select it on the list, and then click Delete.
Name:
Array Indexes:
Port direction
in
inout
out
gates2
Set TimeScale
New
Delete
Type...
< Back
Finish
Cancel
Click New.
Type a.

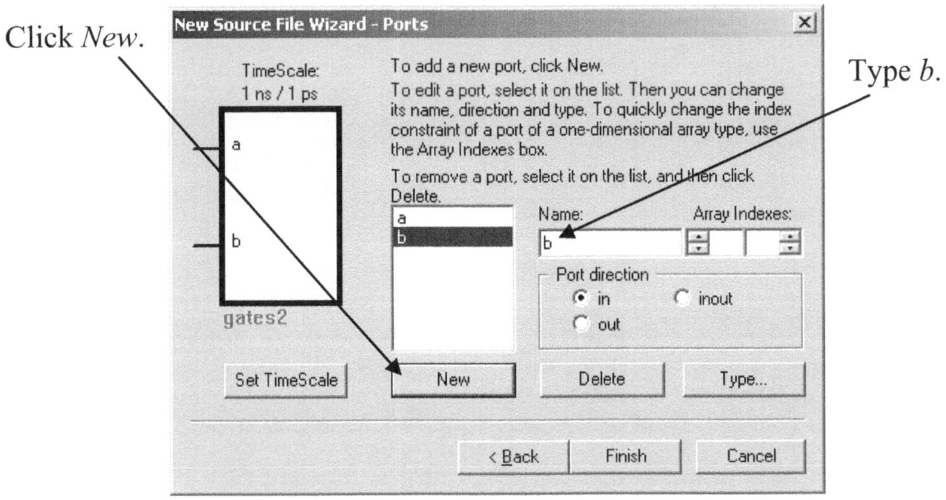
New Source File Wizard - Ports
Click New.
Type b.
Set TimeScale
New
Delete
Type...
< Back
Finish
Cancel

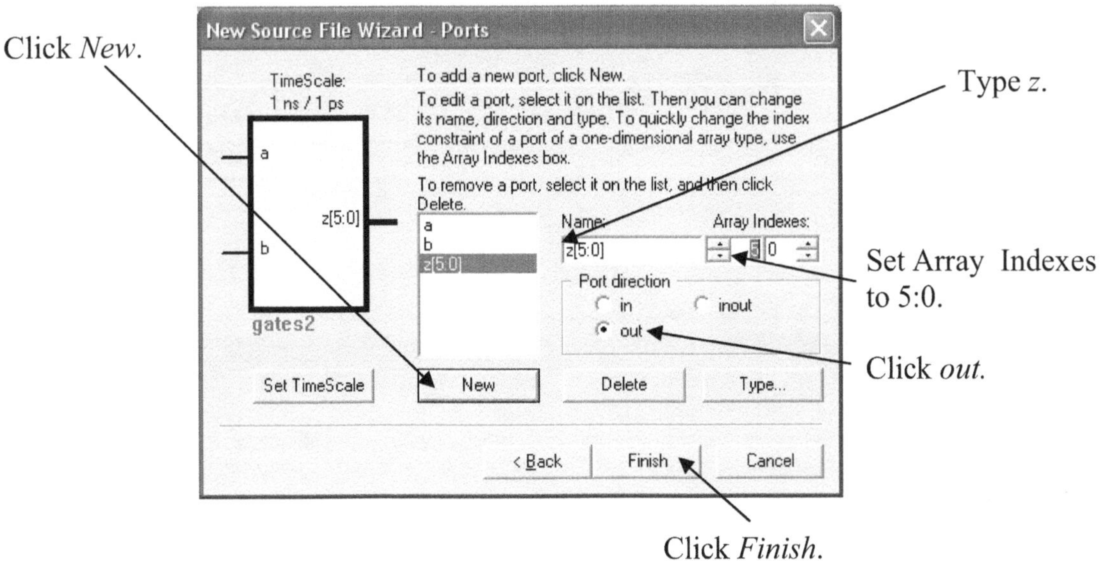
New Source File Wizard - Ports
z[5:0]
Click New.
Type z.
Set Array Indexes to 5:0.
Click out.
Click Finish.

This will generate a VHDL template with the input and output signals filled in. Delete all the comments and replace them with the single comment

```
-- Example 1: 2-input gates
```

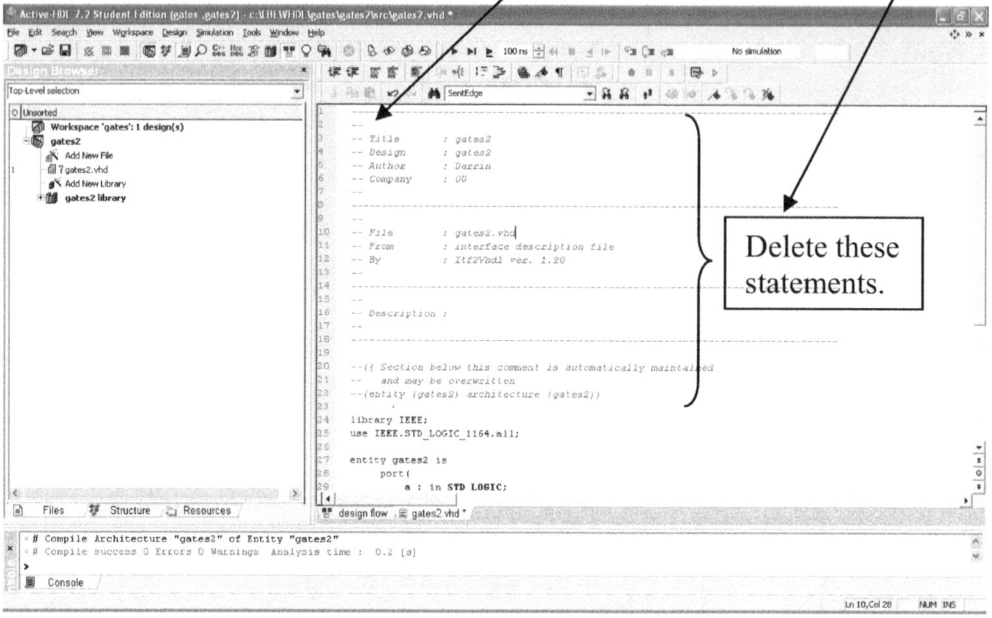

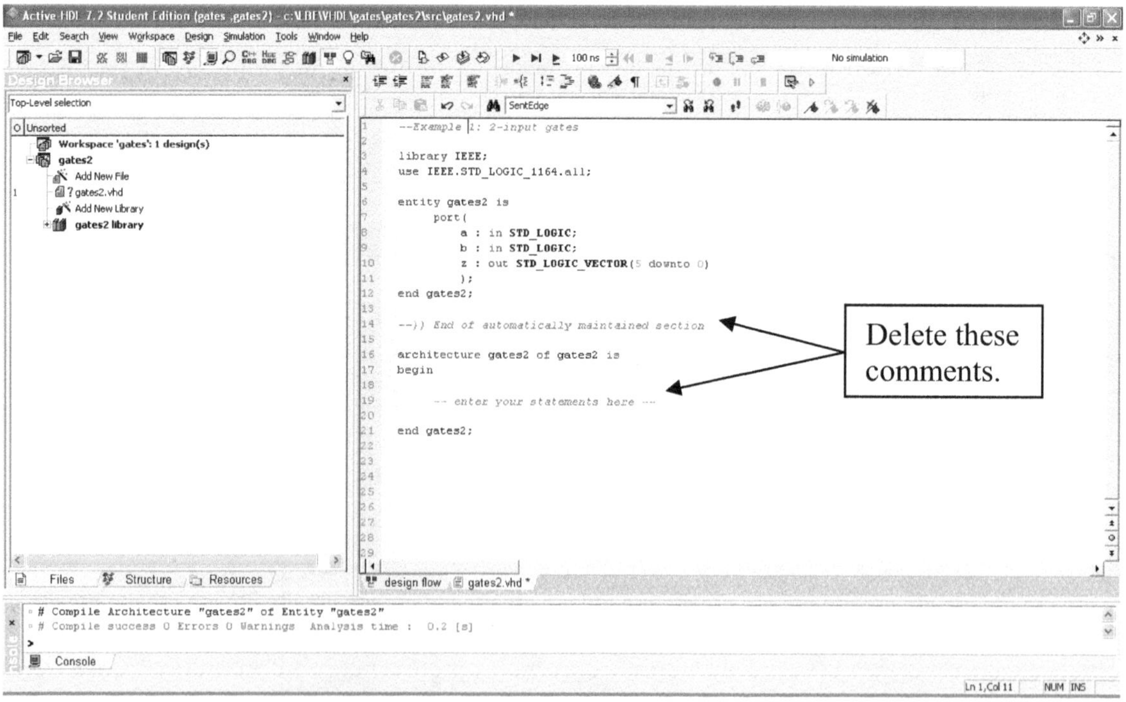

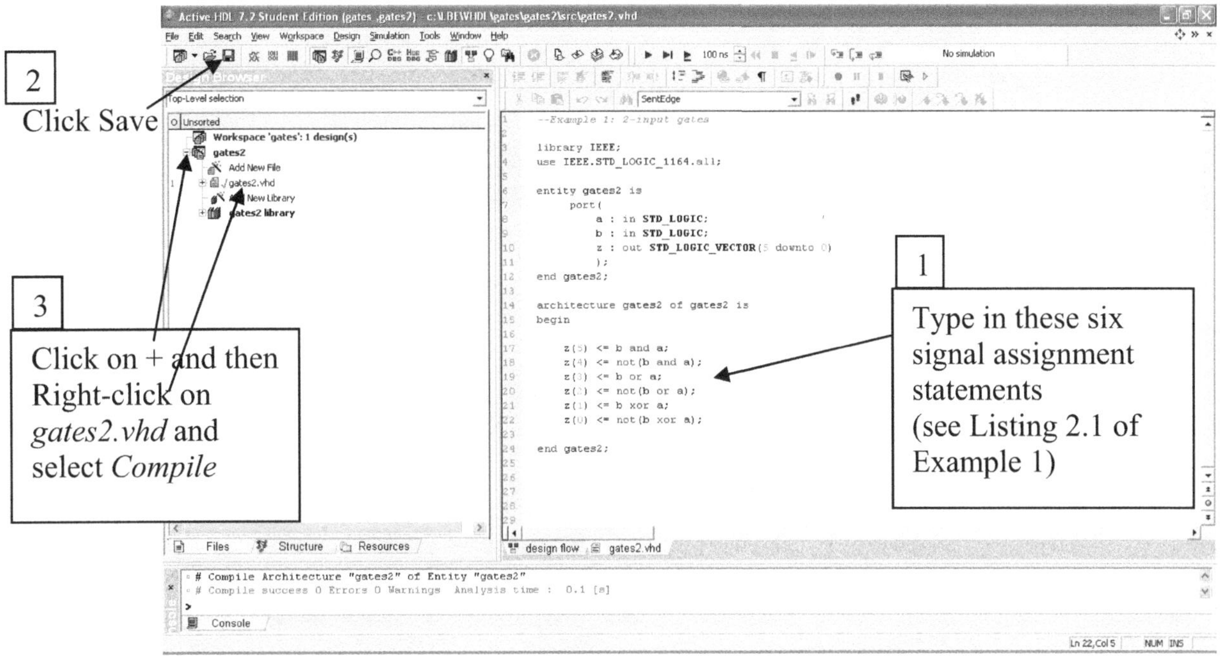

Part 8: Simulation – gates2

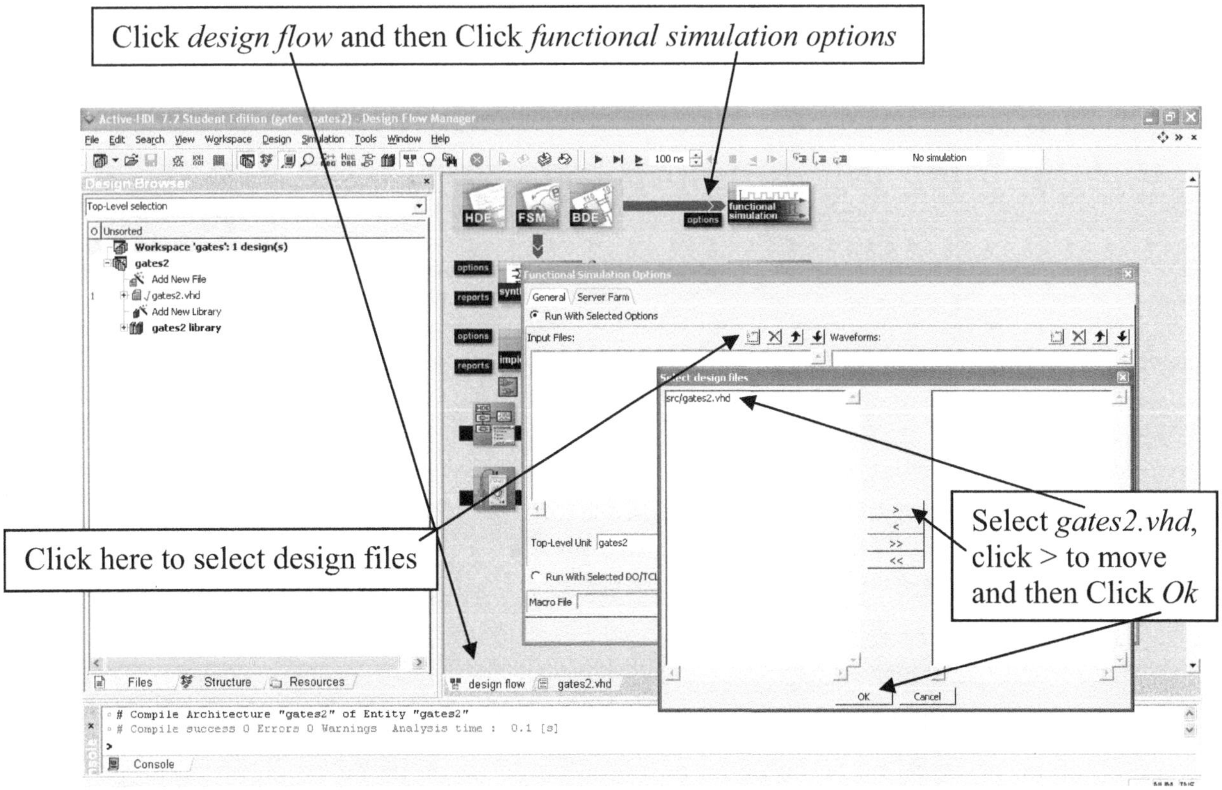

Click *Choose*, select *gates2* as the top-level design, and click *Add*.

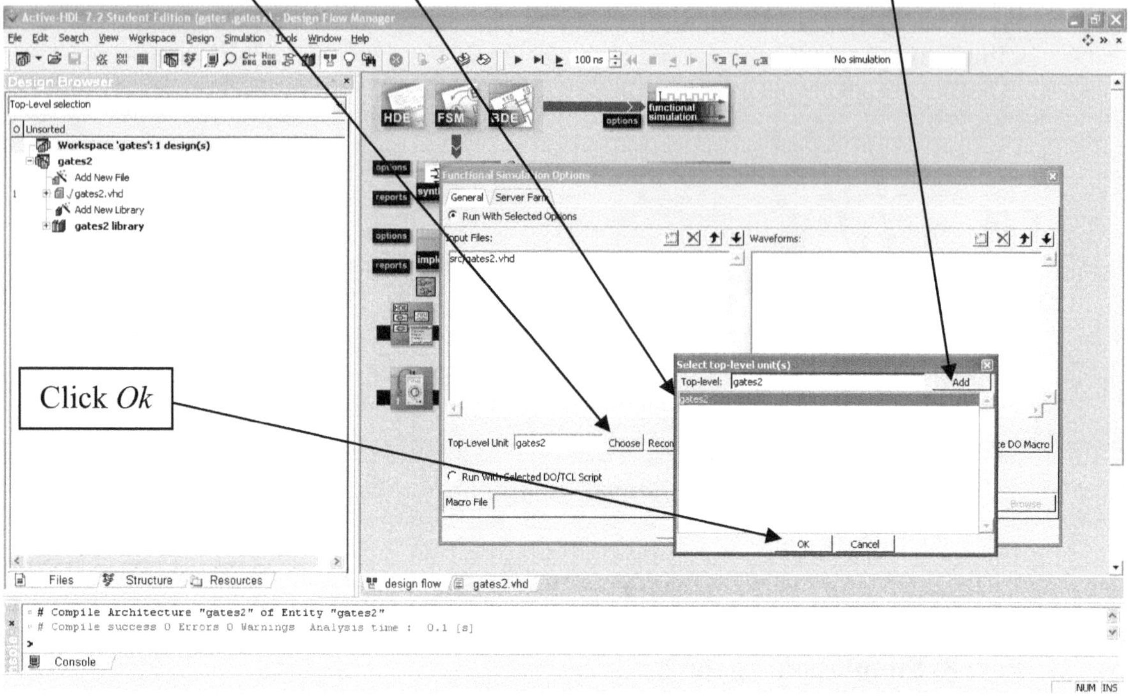

Click *Use Default Waveform*

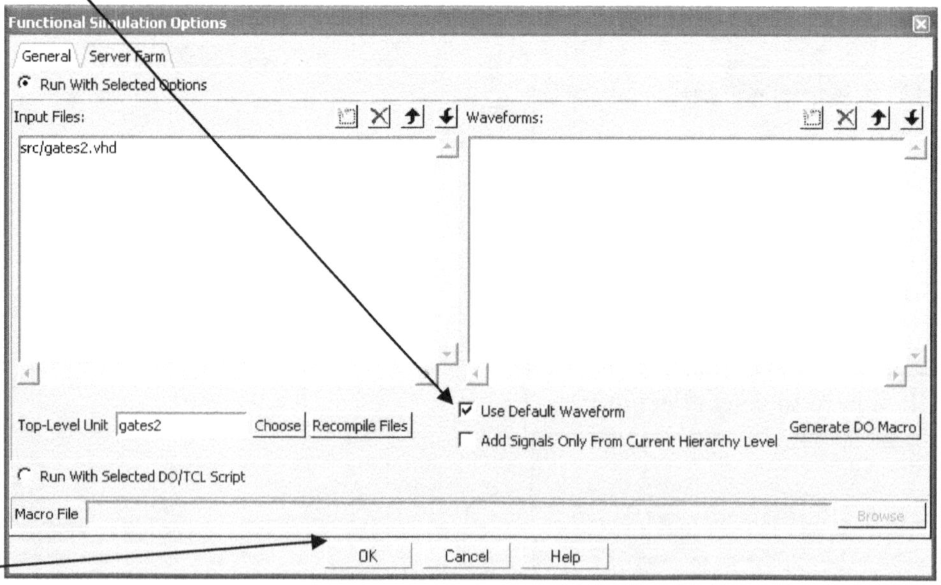

Click *Ok*

Click *functional simulation*

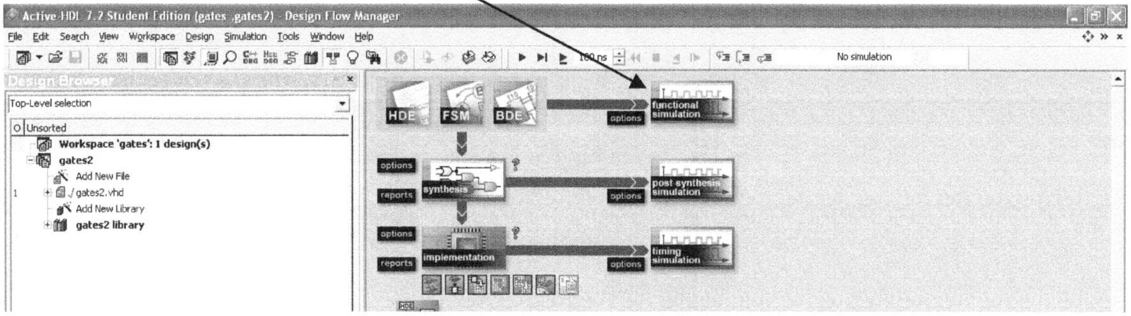

The waveform window will automatically come up with the simulation already initialized. Make sure the order is *a*, *b*, *z* (grab and drag if necessary). Right-click on *a* and select *Stimulators*.

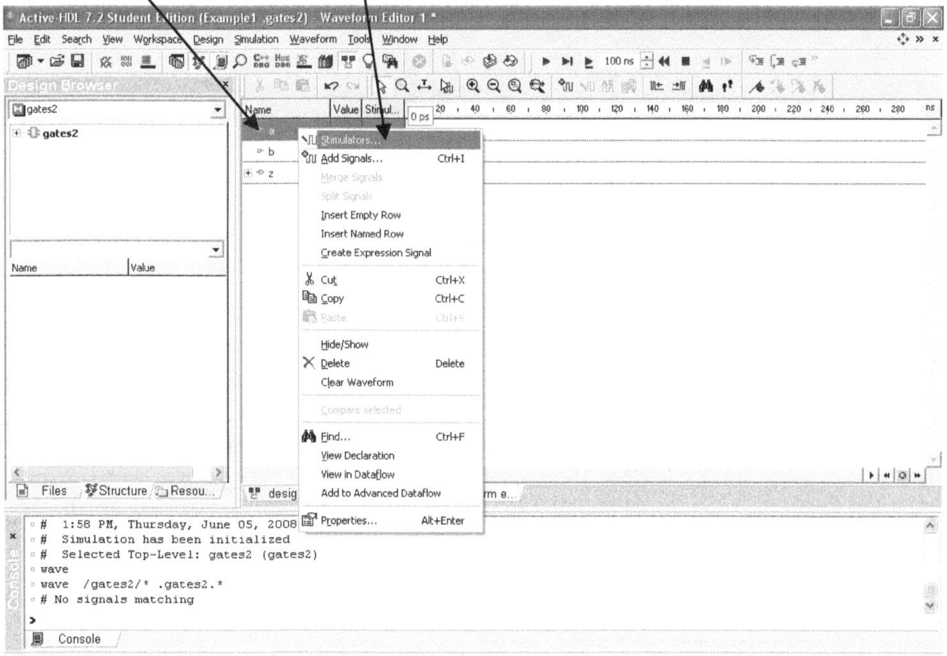

Select *Clock* and set Frequency to 25 MHz

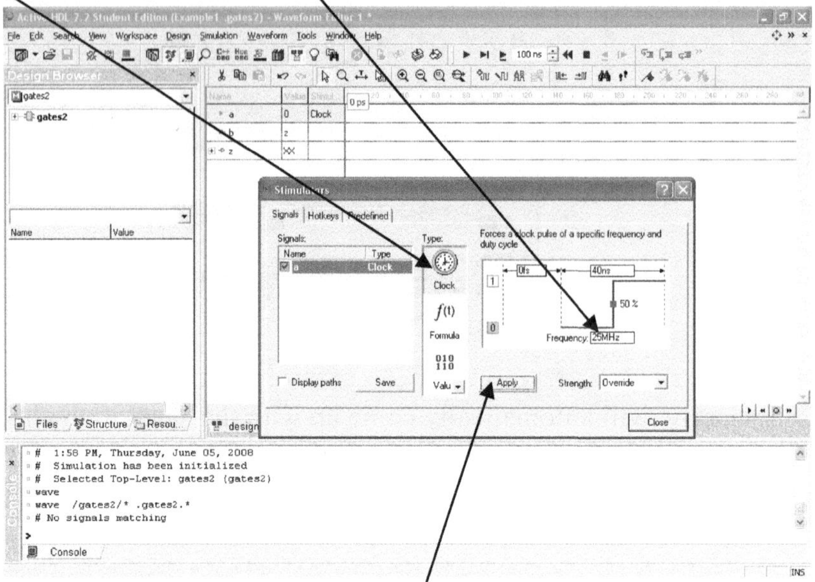

Click *Apply*

Click on *b*, select *Clock* and set Frequency to 50 MHz

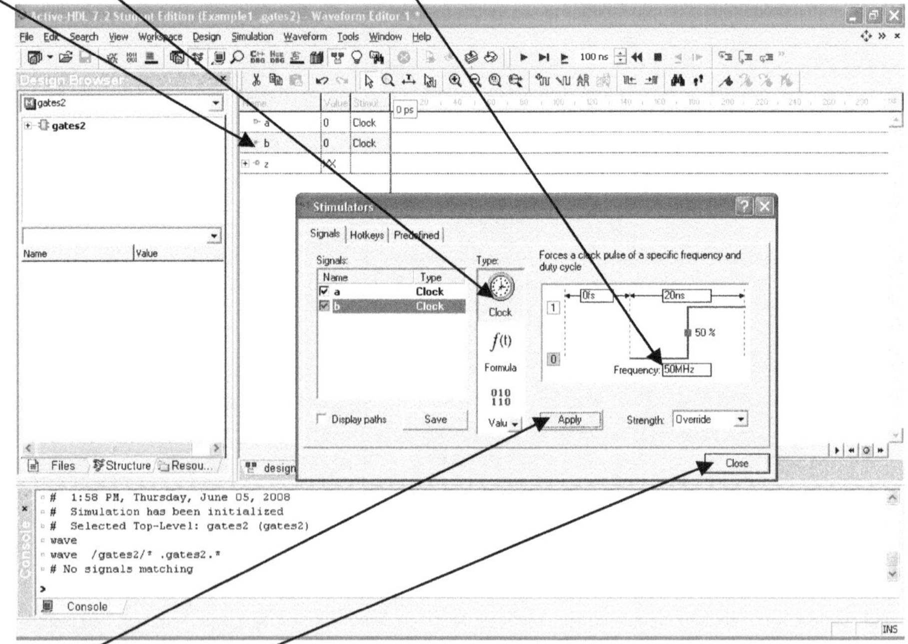

Click *Apply*

Click *Close*

Set simulation time to 50 ns

Click here to run simulation

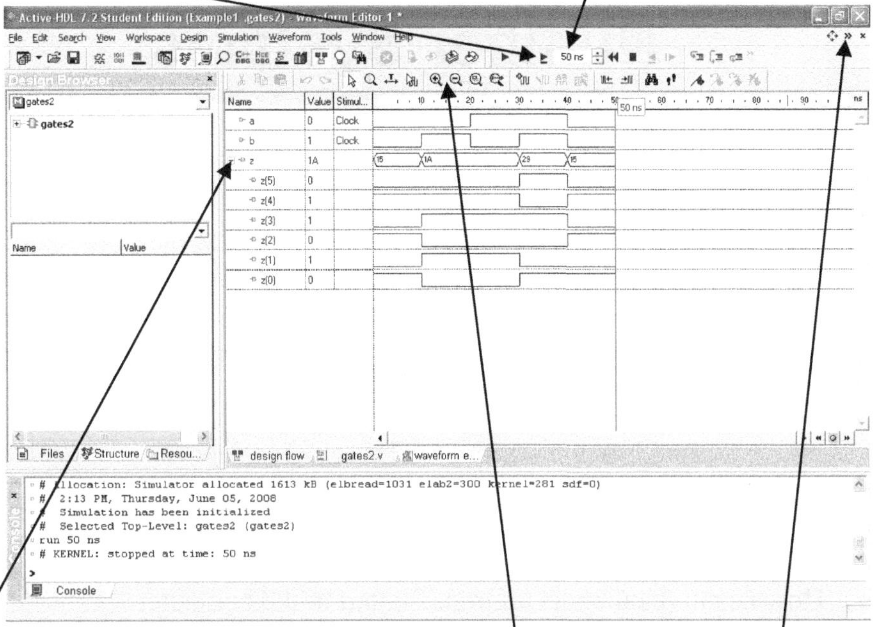

Click + sign to show all elements of *z*.

Study the waveforms for various magnifications.

To print out this waveform you can detach it by clicking >> here and then press *Alt Prnt Scrn* to copy it to the clipboard. Then paste it in a *.doc* file and print.

Appendix B

Number Systems

Data inside a computer are represented by *binary digits* or *bits*. The logical values of these binary digits are denoted by 0 and 1, while the corresponding physical values can be any two-state property such as a high (5 volts) or low (0 volts) voltage or two different directions of magnetization. It is therefore convenient to represent numbers inside the computer in a binary number system. Hexadecimal and octal numbers are often used as a shorthand for representing binary numbers.

In this appendix you will learn:

- How to count in binary and hexadecimal
- How integers and fractional numbers are represented in any base
- How to convert numbers from one base to another
- How negative numbers are represented in the computer

B.1 Counting in Binary and Hexadecimal

Consider a box containing one marble. If the marble is in the box, we will say that the box is *full* and associate the digit 1 with the box. If we take the marble out of the box, the box will be empty, and we will then associate the digit 0 with the box. The two binary digits 0 and 1 are called *bits and* with one bit we can count from zero (box empty) to one (box full) as shown in Fig. B.1.

Figure B.1 You can count from 0 to 1 with 1 bit.

Consider now a second box that can also be full (1) or empty (0). However, when this box is full, it will contain *two* marbles as shown in Fig. B.2. With these two boxes (2 bits) we can count from zero to three, as shown in Fig. B.3. Note that the value of each 2-bit binary number shown in Fig. B.3 is equal to the total number of marbles in the two boxes.

Figure B.2 This box can contain either two marbles (full) or no marbles.

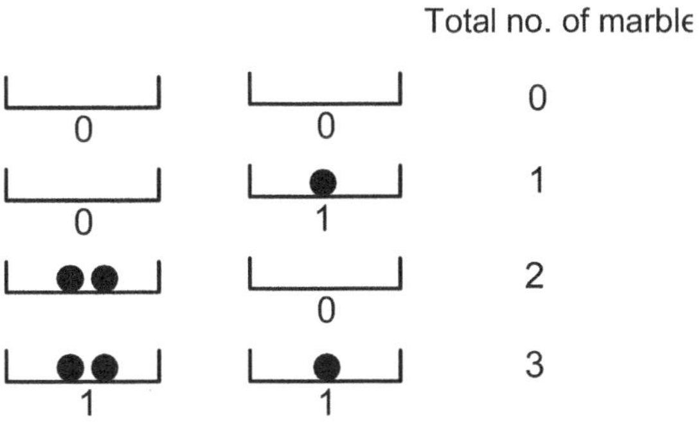

Figure B.3 You can count from 0 to 3 with two bits

We can add a third bit to the binary number by adding a third box that is full (bit = 1) when it contains four marbles and is empty (bit = 0) when it contains no marbles. It must be either full (bit = 1) or empty (bit = 0). With this third box (3 bits), we can count from 0 to 7, as shown in Fig. B.4.

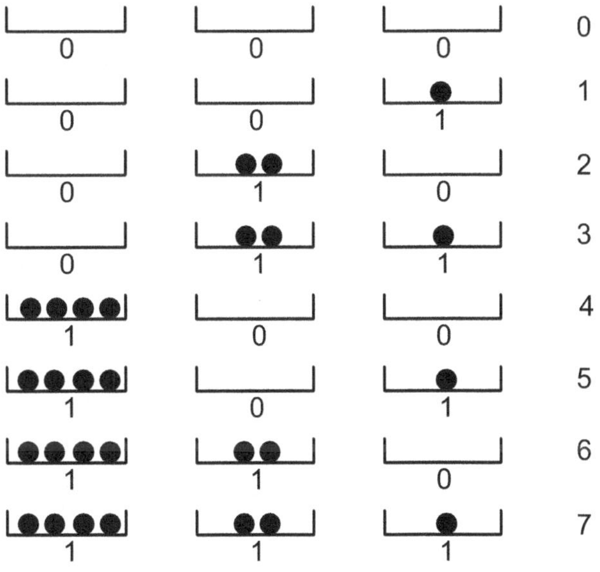

Figure B.4 You can count from 0 to 7 with 3 bits.

If you want to count beyond 7, you must add another box. How many marbles should this fourth box contain when it is full (bit = 1)? It should be clear that this box must contain eight marbles. The binary number 8 would then be written as 1000. Remember that a 1 in a binary number means that the corresponding box is full of marbles, and the number of marbles that constitutes a full box varies as 1, 2, 4, 8, starting at the right. This means that with 4 bits we can count from 0 to 15, as shown in Fig. B.5.

It is convenient to represent the total number of marbles in the four boxes represented by the 4-bit binary numbers shown in Fig. B.5 by a single digit. We call this a *hexadecimal* digit, and the 16 hexadecimal digits are shown in the right column in Fig. B.5. The hexadecimal digits 0 to 9 are the same as the decimal digits 0 to 9. However, the decimal numbers 10 to 15 are represented by the hexadecimal digits A to F. Thus, for example, the hexadecimal digit D is equivalent to the decimal number 13.

No. of marbles in each full box (bit = 1)				Total no. of marbles	Hex digit
8	4	2	1		
0	0	0	0	0	0
0	0	0	1	1	1
0	0	1	0	2	2
0	0	1	1	3	3
0	1	0	0	4	4
0	1	0	1	5	5
0	1	1	0	6	6
0	1	1	1	7	7
1	0	0	0	8	8
1	0	0	1	9	9
1	0	1	0	10	A
1	0	1	1	11	B
1	1	0	0	12	C
1	1	0	1	13	D
1	1	1	0	14	E
1	1	1	1	15	F

Figure B.5 You can count from 0 to 15 with 4 bits.

To count beyond 15 in binary, you must add more boxes. Each full box you add must contain twice as many marbles as the previous full box. With 8 bits you can count from 0 to 255. A few examples are shown in Fig. B.6. The decimal number that corresponds to a given binary number is equal to the total number of marbles in all the boxes. To find this number, just add up all the marbles in the full boxes (the ones with binary digits equal to 1).

No. of marbles in each full box (bit = 1)								Total no. of marbles
128	64	32	16	8	4	2	1	
0	0	1	1	0	1	0	0	52
1	0	1	0	0	0	1	1	163
1	1	1	1	1	1	1	1	255

Figure B.6 You can count from 0 to 255 with 8 bits.

As the length of a binary number increases, it becomes more cumbersome to work with. We then use the corresponding hexadecimal number as a shorthand method of representing the binary number. This is very easy to do. You just divide the binary number into groups of 4 bits starting at the right and then represent each 4-bit group by its corresponding hexadecimal digit given in Fig. B.5. For example, the binary number

1001	1010
9	A

is equivalent to the hexadecimal number $9A. We will often use the dollar sign $ preceding a number to indicate a hexadecimal number. You should verify that the total number of marbles represented by this binary number is 154. However, instead of counting the marbles in the *binary boxes* you can count the marbles in *hexadecimal boxes* where the first box contains A x 1 = 10 marbles and the second box contains 9 x 16 = 144 marbles. Therefore, the total number of marbles is equal to 144 + 10 = 154.

A third hexadecimal box would contain a multiple of 16^2 = 256 marbles, and a fourth hexadecimal number would contain a multiple of 16^3 = 4,096 marbles. As an example, the 16-bit binary number

```
1000|0111|1100|1001
  8 |  7 |  C |  9
```

is equivalent to the decimal number 34,761 (that is, it represents 34,761 marbles). This can be seen by expanding the hexadecimal number as follows:

```
8 x 16^3  =  8 x 4,096 =  32,768
7 x 16^2  =  7 x   256 =   1,792
C x 16^1  = 12 x    16 =     192
9 x 16^0  =  9 x     1 =       9
                         -------
                          34,761
```

Table B.1 will allow you to conveniently convert hexadecimal numbers of up to four digits to their decimal equivalent. Note, for example, how the four terms in the conversion of $87C9 can be read directly from the table.

Table B.1
Hexadecimal and Decimal Conversion

15 Byte			8	7 Byte			0
15 Char	12	11 Char	8	7 Char	4	3 Char	0
Hex	Dec	Hex	Dec	Hex	Dec	Hex	Dec
0	0	0	0	0	0	0	0
1	4,096	1	256	1	16	1	1
2	8,192	2	512	2	32	2	2
3	12,288	3	768	3	48	3	3
4	16,384	4	1,024	4	64	4	4
5	20,480	5	1,280	5	80	5	5
6	24,576	6	1,536	6	96	6	6
7	28,672	7	1,792	7	112	7	7
8	32,768	8	2,048	8	128	8	8
9	36,864	9	2,304	9	144	9	9
A	40,960	A	2,560	A	160	A	10
B	45,056	B	2,816	B	176	B	11
C	49,152	C	3,072	C	192	C	12
D	53,248	D	3,328	D	208	D	13
E	57,344	E	3,584	E	224	E	14
F	61,440	F	3,840	F	240	F	15

B.2 Positional Notation

Binary numbers are numbers to the base 2 and hexadecimal numbers are numbers to the base 16. An integer number N can be written in any base b using the following positional notation:

$$N = P_4P_3P_2P_1P_0 = P_4b^4 + P_3b^3 + P_2b^2 + P_1b^1 + P_0b^0$$

where the number always starts with the least significant digit on the right.

For example, the decimal number 584 is a base 10 number and can be expressed as

$$\begin{aligned} 584_{10} &= 5x10^2 + 8x10^1 + 4x10^0 \\ &= 500 + 80 + 4 \\ &= 584_{10} \end{aligned}$$

A number to the base b must have b different digits. Thus, decimal numbers (base 10) use the 10 digits 0 to 9.

A binary number is a base 2 number and therefore uses only the two digits 0 and 1. For example, the binary number 110100 is the base 2 number

$$\begin{aligned} 110100_2 &= 1x2^5 + 1x2^4 + 0x2^3 + 1x2^2 + 0x2^1 + 0x2^0 \\ &= 32 + 16 + 0 + 4 + 0 + 0 \\ &= 52_{10} \end{aligned}$$

This is the same as the first example in Fig. B.6 where the total number of marbles is 52 (32 + 16 + 4).

A hexadecimal number is a base 16 number and therefore needs 16 different digits to represent the number. We use the ten digits 0 to 9 plus the six letters A to F as shown in Fig. B.5. For example, the hexadecimal number 3AF can be written as the base 16 number

$$\begin{aligned} 3AF_{16} &= 3x16^2 + Ax16^1 + Fx16^0 \\ &= 3x256 + 10x16 + 15x1 \\ &= 768 + 160 + 15 \\ &= 943_{10} \end{aligned}$$

Microcomputers move data around in groups of 8-bit binary bytes. Therefore, it is natural to describe the data in the computer as binary, or base 2, numbers. As we have seen, this is simplified by using hexadecimal numbers where each hex digit represents 4 binary digits. Some older computers represented binary numbers in groups of 3 bits rather than 4. The resulting number is an *octal*, or base 8, number. Octal numbers use only the 8 digits 0 to 7. For example, the octal number 457 can be written as the base 8 number

$$\begin{aligned} 457_8 &= 4x8^2 + 5x8^1 + 8^0 \\ &= 256 + 40 + 7 \\ &= 303_{10} \end{aligned}$$

B.3 Fractional Numbers

The positional notation given in Section B.2 for integer numbers can be generalized for numbers involving fractions as follows:

$$N = \ldots P_2 P_1 P_0 . P_{-1} P_{-2} P_{-3} \ldots = \ldots + P_2 b^2 + P_1 b^1 + P_0 b^0 + P_{-1} b^{-1} + P_{-2} b^{-2} + P_{-3} b^{-3} + \ldots$$

As an example, consider the base 10 number 375.17. Using the above definition, this is equal to

$$\begin{aligned} N &= 3 \times 10^2 + 7 \times 10^1 + 5 \times 10^0 + 1 \times 10^{-1} + 7 \times 10^{-2} \\ &= 300 + 70 + 5 + 0.1 + 0.07 \\ &= 375.17 \end{aligned}$$

In this case the radix, or base, is 10 and the radix point (decimal point) separates the integer part of the number from the fractional part.

Consider now the binary number 1101.11. This is equivalent to what decimal number? Using the above definition, we can write

$$\begin{aligned} 1101.11_2 &= 1\text{x}2^3 + 1\text{x}2^2 + 0\text{x}2^1 + 1\text{x}2^0 + 1\text{x}2^{-1} + 1\text{x}2^{-2} \\ &= 8 + 4 + 0 + 1 + \frac{1}{2} + \frac{1}{4} \\ &= 13.75_{10} \end{aligned}$$

Following the same technique, we can write the hexadecimal number 1AB.6 as

$$\begin{aligned} 1\text{AB}.6_{16} &= 1\text{x}16^2 + 10\text{x}16^1 + 11\text{x}16^0 + 6\text{x}16^{-1} \\ &= 256 + 160 + 11 + \frac{6}{16} \\ &= 427.375_{10} \end{aligned}$$

As a final example consider the octal number 173.25. We can find the equivalent decimal number by expanding the octal number as follows.

$$\begin{aligned} 173.25_8 &= 1\text{x}8^2 + 7\text{x}8^1 + 3\text{x}8^0 + 2\text{x}8^{-1} + 5\text{x}8^{-2} \\ &= 64 + 56 + 3 + \frac{2}{8} + \frac{5}{64} \\ &= 123.328125_{10} \end{aligned}$$

The examples in this section show how you can convert a number in any base to a decimal number. In the following section we will look at how to convert a decimal number to any other base and how to convert among binary, hexadecimal, and octal.

B.4 Number System Conversions

In the previous section you saw how you can convert a number in any base to its decimal equivalent by expanding the number using the definition of the positional notation of the number. For a hexadecimal number containing a maximum of four hex digits, it is easy to use Table B.1 to find the conversion by simply adding the corresponding decimal value from each of the four columns. Note that the entries in the four columns of Table B.1 are simply the hex digits multiplied by 16^3, 16^2, 16^1, and 16^0 respectively.

If you don't have Table B.1 (or a calculator that converts hex numbers to decimal), you can use the following shortcut to convert a hex integer to decimal. To convert the hexadecimal number $A7_{16}$ to decimal, multiply A x 16 and add 7. For longer hexadecimal numbers, start with the leftmost digit (the most significant), multiply it by 16, and add the next hex digit. Multiply this result by 16 and add the next hex digit. Continue this process until you have added the rightmost digit. For example, to convert $87C9_{16}$ to decimal, you can do this:

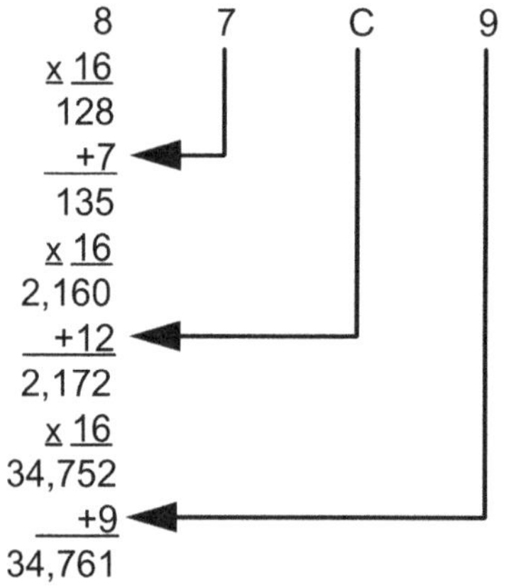

Therefore, $87C9_{16} = 34{,}761_{10}$. This technique will work for any base. You just multiply by the current base, rather than 16, in each step of the process.

Binary <-> Hex

Converting a binary number to hex is trivial. You simply partition the binary number in groups of 4 bits, starting at the radix point, and read the hex digits by inspection using the hex digit definitions in Fig. B.5. For example, the binary number 11010101000.1111010111 can be partitioned as follows:

0110	1010	1000	.	1111	0101	1100
6	A	8	.	F	5	C

Therefore, $11010101000.1111010111_2 = 6A8.F5C_{16}$. Note that leading zeros can be added to the integer part of the binary number, and trailing zeros can be added to the fractional part to produce a 4-bit hex digit.

Going from hex to binary is just as easy. You just write down the 4 binary digits corresponding to each hex digit by inspection (using the table in Fig. B.5).

Binary ≺–≻ Octal

Converting a binary number to octal is just as easy as converting it to hex. In this case you just partition the binary number in groups of 3 bits rather than 4 and read the octal digits (0 to 7) by inspection. Again the grouping is done starting at the radix point. Using as an example, the same binary number 11010101000.1111010111 that we just converted to hex we would convert it to octal as follows:

011	010	101	000	.	111	101	011	100
3	2	5	0	.	7	5	3	4

Therefore, $11010101000.1111010111_2 = 3250.7534_8$. Note again that leading zeros must be added to the integer part of the binary number, and trailing zeros must be added to the fractional part to produce 3-bit octal digits.

You reverse the process to go from octal to binary. Just write down the 3 binary digits corresponding to each octal digit by inspection.

Hex ≺–≻ Octal

When converting from hex to octal or from octal to hex, it is easiest to go through binary. Thus, for example, to convert $6A8.F5C_{16}$ to octal, you would first convert it to the binary number 11010101000.1111010111_2 by inspection, as shown in the example above. Then you would convert this binary number to 3250.7534_8, as we just did in the previous example.

Decimal to Hex

Suppose you want to convert a decimal integer 167 to its hexadecimal equivalent. The easiest way to figure this out is to look at Table B.1. The closest decimal value in this table that does not exceed 167 is 160 in the second column from the right. This corresponds to the hexadecimal digit A as the second digit from the right ($A \times 16^1 = 10 \times 16 = 160$). To find the hexadecimal digit to use in the rightmost position, subtract 160 from 167. Thus the decimal number 167_{10} is equivalent to the hexadecimal number $A7_{16}$. What binary number is this?

How can you convert a decimal integer to hexadecimal if you don't have Table B.1 around? Here's a shortcut. Divide the decimal number by 16 and keep track of the remainder. Keep dividing the results by 16 and writing down the remainders at each step until the result is zero. The equivalent hexadecimal number is all the remainders read backward. For example, this is how to convert the decimal number 167_{10} to hexadecimal:

```
167/16 =  10 with remainder    7
 10/16 =   0 with remainder   10 = A
                               ↑
             read backward ────┘
```

$\therefore\ 167_{10} = A7_{16}$

Here's the example we gave at the beginning of this section.

$34{,}761_{10} = ?_{16}$

```
34,761/16 = 2,172   with remainder  9
 2,172/16 =   135   with remainder 12 = C
   135/16 =     8   with remainder  7
     8/16 =     0   with remainder  8
```

read up ↑

Therefore, $34{,}761_{10} = 87C9_{16}$. Again, this technique will work for converting a decimal integer to any base. You just divide by the base, keep track of the remainders, and read up.

When converting a decimal number containing a fractional part, you divide the problem into two parts. First, convert the integer part using the technique just described. Then you can use the following rule to convert the fractional part: Multiply the fractional part by the base, keep track of the integer part, and read down. As an example, suppose you want to convert the decimal number 3901.78125_{10} to its hexadecimal equivalent. You would first convert the integer part by dividing by the base, keeping track of the remainder, and reading up:

```
3901/16 = 243   with remainder 13 = D
 243/16 =  15   with remainder       3
  15/16 =   0   with remainder 15 = F
```

read up ↑

Therefore, $3901_{10} = F3D_{16}$. To convert the fractional part, multiply by the base, keep track of the integer part, and read down:

read down ↓

```
0.78125 x 16 = 12.5   integer part = 12 = C
    0.5 x 16 =  8.0   integer part = 8
```

Therefore, $0.78125_{10} = 0.C8_{16}$. Combining the integer and fractional parts, we have found that $3901.78125_{10} = F3D.C8_{16}$.

This rule for converting the fractional part of a decimal number will work for any base. Sometimes the remainder may never become zero and you will have a continuing fraction. This means that there is no exact conversion of the decimal fraction. For example, suppose you want to represent the decimal value 0.1_{10} as a binary number. Following our rule, we would write

```
                                   read down
0.1 x 2 = 0.2      integer part = 0
0.2 x 2 = 0.4      integer part = 0
0.4 x 2 = 0.8      integer part = 0
0.8 x 2 = 1.6      integer part = 1
0.6 x 2 = 1.2      integer part = 1
0.2 x 2 = 0.4      integer part = 0
0.4 x 2 = 0.8      integer part = 0
0.8 x 2 = 1.6      integer part = 1
0.6 x 2 = 1.2      integer part = 1
```

It is clear that the remainder will never go to zero and that the pattern 0011 will go on forever. Thus, 0.1_{10} can only be approximated as

$$0.1_{10} = 0.000110011..._2$$

This means that 0.1_{10} cannot be represented exactly in a computer as a binary number of any size!

B.5 Negative Numbers

An 8-bit binary number can represent one of 256 (2^8) possible values between 0 and 255. However, we also need to represent negative numbers. The leftmost bit in a binary number is the *sign bit.* If this bit is zero, the number is positive; if this bit is one, the number is negative. However, in the 8086, (and in most computers today), when the most significant bit is one, the magnitude of the negative number is *not* given by the binary value of the remaining bits in the number. Rather a two's complement representation of negative numbers is used. The reason for this is that the same circuitry, an adder, can be used for both addition and subtraction.

The idea of being able to subtract by adding can be seen by an example using decimal (base 10) numbers. Suppose you want to subtract 35 from 73. The answer is 38. You can obtain this result by taking the 10's complement of 35 (this is what you have to add to 35 to get 100; that is, 65) and adding it to 73. The result is 138 as shown in Fig. B.7. If you ignore the leading 1 (the carry) then the remaining value, 38, is the correct answer.

```
 73                          73
-35     10's complement     +65
---                         ---
 38                         138

              Ignore carry
```

Figure B.7 Decimal subtraction can be done by taking the 10's complement of the subtrahend and adding

In binary arithmetic, negative numbers are stored in their two's complement form. You can find the two's complement of a binary number in several ways. Note that the 10's complement of 35 can be found by subtracting 35 from 99 (this gives the 9's complement) and then adding 1. That is,

```
 99
-35
---
 64
+ 1
---
 65
```

The two's complement of the 8-bit binary number 01001101 is the 8-bit binary number you must add to this number to obtain 100000000. You can find it by subtracting the number from 11111111 and adding 1. Note that subtracting an 8-bit binary number from 11111111 (called the one's complement) is equivalent to complementing each bit in the byte; that is, each 1 is changed to a 0, and each 0 is changed to a 1. Therefore, the one's complement of 01001101 is 10110010 and the two's complement of 01001101 is

```
                      01001101
one's complement   =  10110010
             add             1
                      --------
two's complement   =  10110011
```

There is an easier way to take the two's complement of a binary number. You just start at the rightmost bit and copy down all bits until you have copied down the first 1. Then complement (that is, change from 1 to 0 or 0 to 1) all the remaining bits. For example,

complement remaining bits ↓ ↓ copy this first 1

```
                     0 1 0 0 1 1 0 1
two's complement  =  1 0 1 1 0 0 1 1
```

As a second example,

complement remaining bits ↓ ↓ copy all bits to first 1

```
                     0 1 0 1 1 0 0 0
two's complement  =  1 0 1 0 1 0 0 0
```

Verify that if you add the 8-bit binary numbers given in the examples above to their two's complement value you obtain 100000000.

An 8-bit byte can contain positive values between 00000000 and 01111111; that is, between \$00 and \$7F. This corresponds to decimal values between 0 and 127. A byte in which bit 7 is a 1 is interpreted as a negative number, whose magnitude can be found by taking the two's complement. For example, how is -75_{10} stored in the computer? First, write down the binary or hex value of the number, \$4B, as shown in Fig. B.8. Then take its two's complement. The value -75_{10} is therefore stored in the computer as \$B5.

Note that if you take the two's complement of a positive number between 0 and \$7F the result will always have bit 7 (the most significant bit) set to 1.

	75_{10}	= \$4B	= 01001011
Two's complement =	-75_{10}	= \$B5	= 10110101
Two's complement of	\$B5	= \$4B	= 01001011

Figure B.8 The negative of a binary number is found by taking the two's complement.

Given a negative number (with bit 7 set), you can always find the magnitude of this number by taking the two's complement. For example, the two's complement of \$B5 (-75_{10}) is \$4B ($+75_{10}$), as shown in Fig. B.8.

Note that the two's complement of \$01 is \$FF and the two's complement of \$80 is \$80, as shown in Fig. B.9. This last example shows that signed 8-bit binary numbers "wrap around" at \$80. That is, the largest positive number is \$7F = 127_{10} and the smallest negative number (largest magnitude) is \$80 = -128_{10}. This is shown in Table B.2.

	1_{10}	= \$01	= 00000001
Two's complement =	-1_{10}	= \$FF	= 11111111
	128_{10}	= \$80	= 10000000
Two's complement =	-128_{10}	= \$80	= 10000000

Figure B.9 Negative 8-bit numbers can range between \$FF (-1) and \$80 (-128).

Table B.2
Positive and Negative Binary Numbers

Signed decimal	Hex	Binary	Unsigned decimal
-128	80	10000000	128
-127	81	10000001	129
-126	82	10000010	130
...	...	...	...
...	...	...	...
...	...	...	...
-3	FD	11111101	253
-2	FE	11111110	254
-1	FF	11111111	255
0	00	00000000	0
1	01	00000001	1
2	02	00000010	2
3	03	00000011	3
...	...	...	...
	...	...	...
...		...	...
125	7D	01111101	125
126	7E	01111110	126
127	7F	01111111	127

Table B.2 also shows that the hex values between $80 and $FF can be interpreted *either* as negative numbers between -128 and -1 *or* as positive numbers between 128 and 255. Sometimes you will treat them as negative numbers and sometimes you will treat them as positive values. It is up to you to make sure you know whether a particular byte is being treated as a signed or as an unsigned number.

Whereas bit 7 is the sign bit in an 8-bit byte, bit 15 is the sign bit in a 16-bit word. A 16-bit signed word can have values ranging from $8000 = $-32{,}768_{10}$ to $7FFF = $+32{,}767_{10}$. Similarly, bit 31 is the sign bit in a 32-bit double word. Such a double word can have values ranging from $80000000 = $-2{,}147{,}483{,}648_{10}$ to $7FFFFFFF = $+2{,}147{,}483{,}647_{10}$.

Appendix C

Basic Logic Gates

All digital systems are made from a few basic digital circuits that we call *logic gates*. These circuits perform the basic logic functions that we will describe in this chapter. The physical realization of these logic gates has changed over the years from mechanical relays to electronic vacuum tubes to transistors to integrated circuits containing thousands of transistors.

In this chapter you will learn:

- Definitions of the basic gates in terms of truth tables and logic equations
- DeMorgan's Theorem
- How gates defined in terms of positive and negative logic are related
- How to write a logic equation from a truth table using sum-of-products and product-of-sum designs

C.1 Truth Tables and Logic Equations

All data in a computer are stored as binary digits. These bits can be thought of as the logical values 0 and 1, where a 1 is considered to be *true* and a 0 is considered to be *false*. The actual physical quantities associated with a 0 and a 1 might be a low (0 volts) or high (5 volts) voltage.

A truth table will define the logical outputs (0 or 1) of the gate for all possible logical inputs. In this section we will define the three basic gates, NOT, AND, and OR, by means of their truth tables. We will then use these basic gates to define some additional gates. Using truth tables we will discover the important De Morgan's theorem. We will then consider the possibility of considering 0 to be *true* and 1 to be *false*. This will give us a better insight into the various gates.

The Three Basic Gates

NOT gate. The definition of the NOT gate, or *inverter*, is shown in Fig. C.1. The logic symbol for the inverter has a single input x and a single output y. The value of y is the complement of the input. Thus, as shown in the truth table in Fig. C.1 if x is 0, then y is 1, whereas if x is 1, then y is 0. The NOT gate simply inverts the logic state of the input.

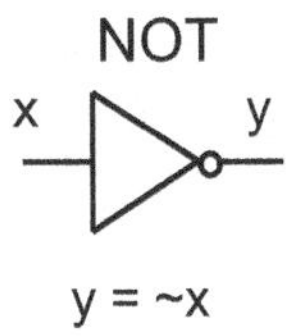

x	y
0	1
1	0

Figure C.1 The NOT gate or inverter.

The equation for the inverter in Fig. C.1 is given as $y = \sim x$. We read this as "*y* equals *NOT x*." This is the symbol used for NOT in Verilog. VHDL uses the word **not** as the negation operator. Thus, in VHDL the logic equation $y = \sim x$ would be written as *y* <= **not** *x* (the use of the assignment operator <= will be discussed later). Because writing out the word **not** is more cumbersome, we will use the Verilog operator ~ for the NOT operation when writing logic equations in this book. The prime, bar, exclamation point, slash, and ¬ are sometimes used to indicate the NOT operation, as in

$$y = x' \qquad y = !x \qquad y = \overline{x} \qquad y = /X \qquad y = \neg X$$

AND Gate. The definition of the AND gate is shown in Fig. C.2. The AND gate logic symbol has two inputs, *x* and *y* and the single output *z*. From the truth table in Fig. C.2 we see that the output *z* of an AND gate is 1 (true or high) only if *both* inputs, *x and y*, are 1 (true or high). The output *z* will be zero if either *x* or *y* or both are zero.

The equation for the AND gate in Fig. C.2 is given as $z = x \,\&\, y$. We read this as "*z* equals *x AND y*." This is the symbol used for AND in Verilog. VHDL uses the word **and** as the AND operator. Thus, in VHDL the logic equation $z = x \,\&\, y$ would be written as *z* <= *x* **and** *y*. Again, for convenience we will use the Verilog notation & for the AND operation when writing logic equations. Other common ways to indicate the AND operation are

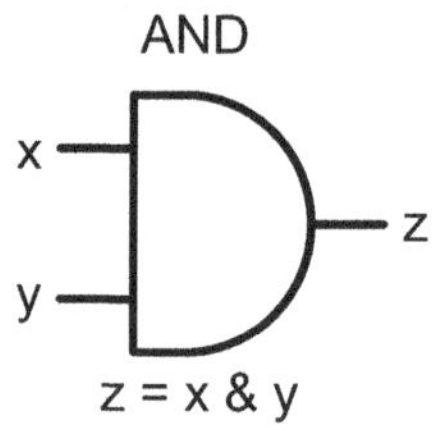

x	y	z
0	0	0
0	1	0
1	0	0
1	1	1

Figure C.2 The AND gate.

$$x \cap y \qquad x * y \qquad xy$$

The last form involving the juxtaposition of *x* and *y* limits you to logic variables containing a single letter. We will be using names for our logic variables in which case *xy* could represent a *single* logic variable.

OR Gate. The definition of the OR gate is shown in Fig. C.3. The OR gate logic symbol has two inputs, *x* and *y*, and the single output *z*. From the truth table in Fig. C.3 we see that the output *z* of an OR gate is 1 (true or high) if *either* input, *x or y*, or both are 1 (true or high). The output *z* will be zero only if both *x* and *y* are zero.

The equation for the OR gate in Fig. C.3 is given as $z = x \mid y$. We read this as "*z* equals *x OR y*." This is the symbol used for OR in Verilog. VHDL uses the word **or** as the OR operator. Thus, in VHDL the logic equation $z = x \mid y$ would be written as *z* <= *x* **or** *y*. Again, for convenience we will use the Verilog notation | for the OR operation when writing logic equations. Other common ways to indicate the OR operation are

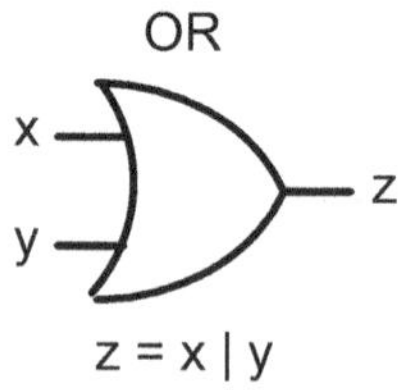

x	y	z
0	0	0
0	1	1
1	0	1
1	1	1

Figure C.3 The OR gate.

$$x + y \qquad x \,\#\, y \qquad x \vee y \qquad x \cup y$$

As surprising as it may seem, all digital systems, including all computers, can be built from only the three basic gates: NOT, AND, and OR. We will begin by showing that three other common gates can be built from our basic three.

Four New Gates

Four new gates, NAND, NOR, Exclusive-OR, and Exclusive NOR can be formed from our three basic gates: NOT, AND, and OR.

NAND Gate. The definition of the NAND gate is shown in Fig. C.4. The logic symbol for a NAND gate is like an AND gate with a small circle (or bubble) on the output. From the truth table in Fig. C.4, we see that the output of a NAND gate is 0 (low) only if both inputs are 1 (high). The NAND gate is equivalent to an AND gate followed by an inverter (NOT-AND), as shown by the two truth tables in Fig. C.4. VHDL uses the work **nand** for the NAND operation.

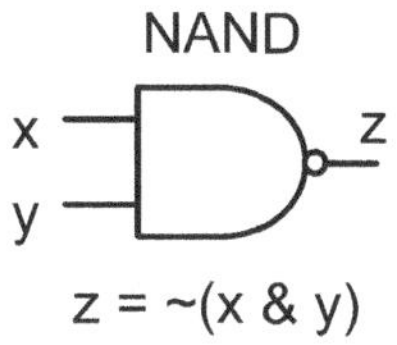

x	y	z
0	0	1
0	1	1
1	0	1
1	1	0

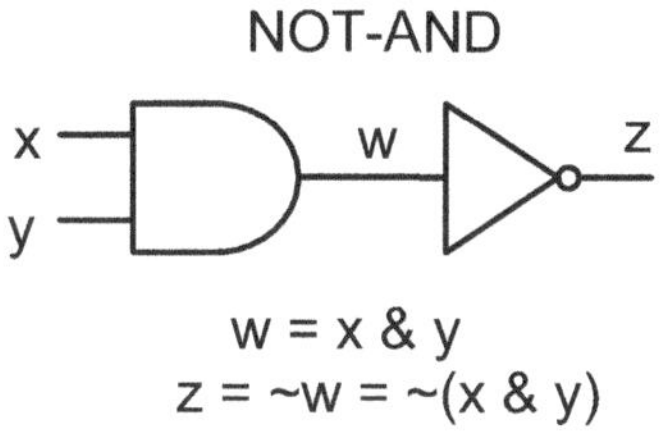

x	y	w	z
0	0	0	1
0	1	0	1
1	0	0	1
1	1	1	0

Figure C.4 The NAND

NOR Gate. The definition of the NOR gate is shown in Fig. C.5. The logic symbol for a NOR gate is like an OR gate with a small circle (or bubble) on the output. From the truth table in Fig. C.5 we see that the output of a NOR gate is 1 (high) only if both inputs are 0 (low). The NOR gate is equivalent to an OR gate followed by an inverter (NOT-OR), as shown by the two truth tables in Fig. C.5. VHDL uses the work **nor** for the NOR operation.

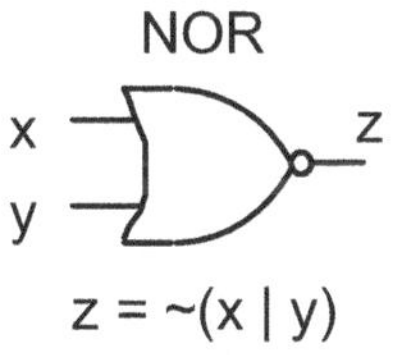

x	y	z
0	0	1
0	1	0
1	0	0
1	1	0

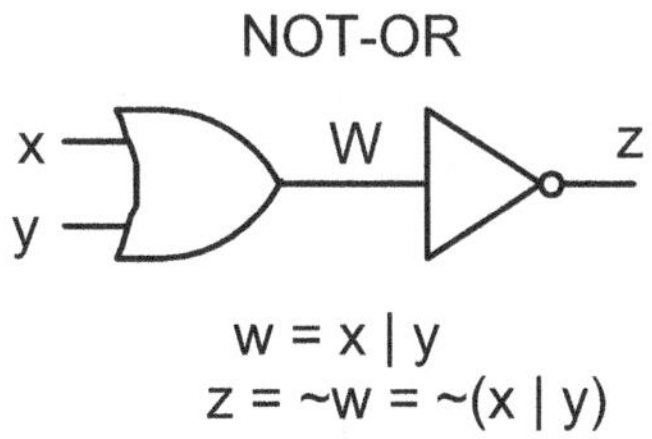

x	y	w	z
0	0	0	1
0	1	1	0
1	0	1	0
1	1	1	0

Figure C.5 The NOR gate.

Exclusive-OR Gate. The definition of the Exclusive-OR, or XOR, gate is shown in Fig. C.6. The XOR gate logic symbol is like an OR gate symbol with an extra curved vertical line on the input. From the truth table in Fig. C.6 we see that the output *z* of an

XOR gate is 1 (true or high) if *either* input, *x or y*, is 1 (true or high), but *not* both. The output *z* will be zero if both *x* and *y* are the same (either both 1 or both 0).

The equation for the XOR gate in Fig. C.6 is given as *z* = *x* ^ *y*. This is the symbol used for XOR in Verilog. VHDL uses the word **xor** as the XOR operator. Thus, in VHDL the logic equation *z* = *x* ^ *y* would be written as *z* <= *x* **xor** *y*. Again, for convenience we will use the Verilog notation ^ for the XOR operation when writing logic equations. Sometimes the symbol ⊕ is used to denote Exclusive-OR.

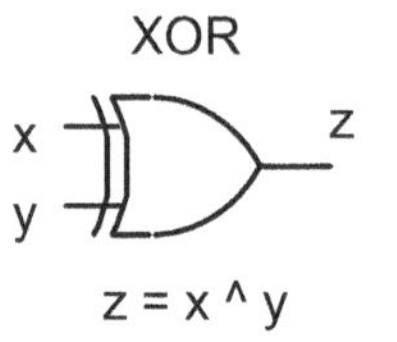

x	y	z
0	0	0
0	1	1
1	0	1
1	1	0

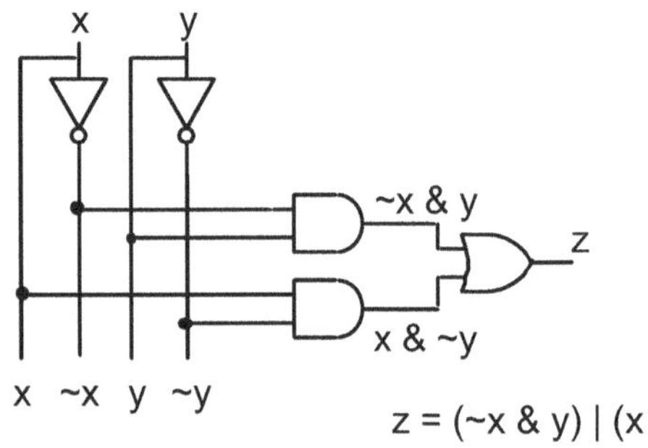

x	y	~x	~y	~x & y	x & ~y	z
0	0	1	1	0	0	0
0	1	1	0	1	0	1
1	0	0	1	0	1	1
1	1	0	0	0	0	0

z = (~x & y) | (x & ~y)

Figure C.6 The Exclusive-OR (XOR) gate.

From Fig. C.6 we see that the XOR gate can be formed with two inverters, two AND gates and an OR gate. Note from this figure and truth table that the Exclusive-OR can be written as

```
z = x ^ y
  = (~x & y) | (x & ~y)
```

Exclusive-NOR Gate. The definition of the Exclusive-NOR, or XNOR, gate is shown in Fig. C.7. The XNOR gate is a NOT-XOR gate and the logic symbol is an XOR gate symbol with a bubble on the output. From the truth table in Fig. C.7 we see that the output *z* of an XNOR gate is 1 (true or high) if both inputs, *x* and *y*, are equal (either both 1 or both 0). The output *z* will be zero if *either* input, *x or y*, is 1 (true or high), but *not* both.

The equation for the XNOR gate in Fig. C.7 is given as *z* = *x* ~^ *y*. This is the symbol used for XNOR in Verilog. VHDL uses the word **xnor** as the XNOR operator. Thus, in VHDL the logic equation *z* = *x* ~^ *y* would be written as *z* <= *x* **xnor** *y*. Again, for convenience we will use the Verilog notation ~^ for the XNOR operation when writing logic equations. Sometimes the symbol ⊙ is used to denote Exclusive-NOR.

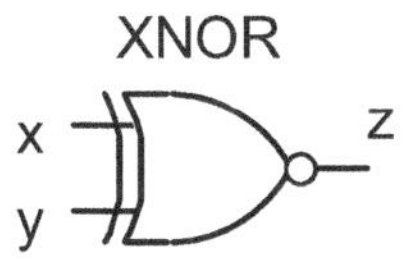

z = ~(x ^ y)
z = x ~^ y

x	y	z
0	0	1
0	1	0
1	0	0
1	1	1

Figure C.7 The Exclusive-NOR (XNOR) gate.

C.2 Positive and Negative Logic: De Morgan's Theorem

In the above examples we considered a 1 to be *true* and a 0 to be *false*. This is called *positive logic*. Another way to interpret our NAND and NOR gates is to think of an output containing the bubble (or small circle) as being *true* when the output is 0. We say that the output is *active low*. This is *negative logic*. Then the NAND gate is just an AND gate in which the output is *true* (0 or active low) only when both inputs are true (1 or active high). Look at the truth table for the NAND gate in Fig. C.4 to see this.

We can even put bubbles on the input to our gates and think of them as having true values when the inputs are 0 or active low (negative logic). Remember that the bubble is equivalent to putting an inverter (NOT gate) there. If we put two bubbles on the input to an OR gate, we get a NAND gate as shown in Fig. C.8. Compare the truth table in Fig. C.8 with that in Fig. C.4 to see that it really is a NAND gate. The OR-type symbol in Fig. C.8 is just an alternate representation of a NAND gate.

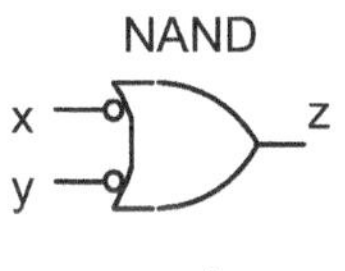

z = ~x | ~y

x	y	z
0	0	1
0	1	1
1	0	1
1	1	0

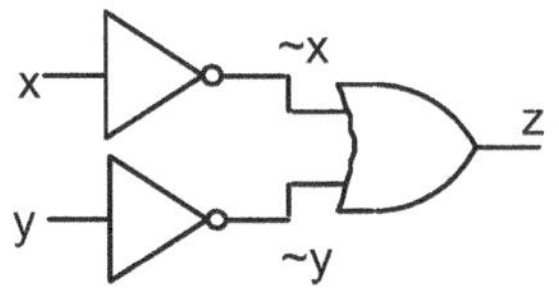

x	y	~x	~y	z
0	0	1	1	1
0	1	1	0	1
1	0	0	1	1
1	1	0	0	0

Figure C.8 Alternate representation of a NAND gate.

Note that in the representation shown in Fig. C.8 we can think of a NAND gate as an OR gate in which the output is *true* (1 or active high) if either or both inputs are *true* (0 or active low). Because the two forms of the NAND gate shown in Figs. C.4 and C.8 are equivalent, that is, they have the same truth table, the two equations for the NAND gate given in the figures must be equal. We can therefore write

```
z = ~x | ~y = ~(x & y)                    (C.1)
```

which is one form of *De Morgan's theorem.*

Let's apply this same idea of thinking of an output containing the bubble as being *true* when the output is 0, or active low, to the NOR gate in Fig. C.5. Then the NOR gate is just an OR gate in which the output is *true* (0 or active low) when either or both inputs are true (1 or active high).

If we put two bubbles on the input to an AND gate, we get a NOR gate, as shown in Fig. C.9. Compare the truth table in Fig. C.9 with that in Fig. C.5 to see that it really is a NOR gate. The AND-type symbol in Fig. C.9 is just an alternate representation of a NOR gate.

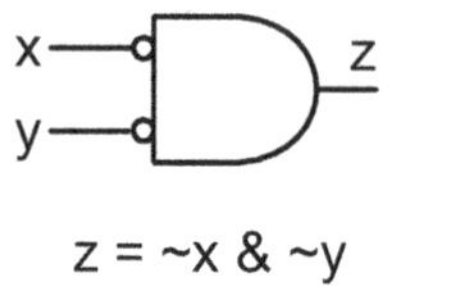

x	y	z
0	0	1
0	1	0
1	0	0
1	1	0

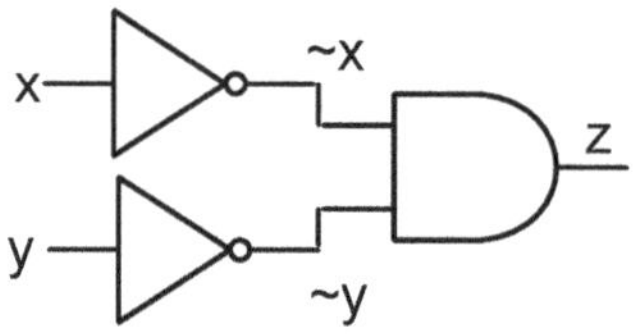

x	y	~x	~y	z
0	0	1	1	1
0	1	1	0	0
1	0	0	1	0
1	1	0	0	0

Figure C.9 An alternate representation of a NOR gate.

Note in the representation shown in Fig. C.9 that we can think of a NOR gate as an AND gate in which the output is *true* (1 or active high) only if both inputs are *true* (0 or active low). Because the two forms of the NOR gate shown in Figs. C.5 and C.9 are equivalent, that is, they have the same truth table, the two equations for the NOR gate given in these figures must be equal. We can therefore write

```
z = ~x & ~y = ~(x | y)                                  (C.2)
```

which is another form of *De Morgan's theorem.*

The symbol for an inverter can also have the bubble on the input, as shown in Fig. C.7. From this figure we see that

```
~~x = x                                                 (C.3)
```

which represents two inverters forming a noninverting buffer.

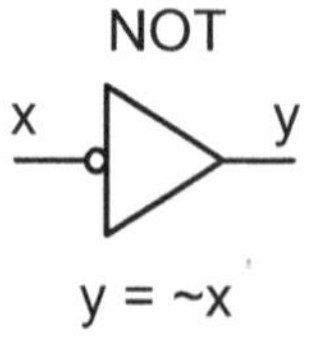

x	y
0	1
1	0

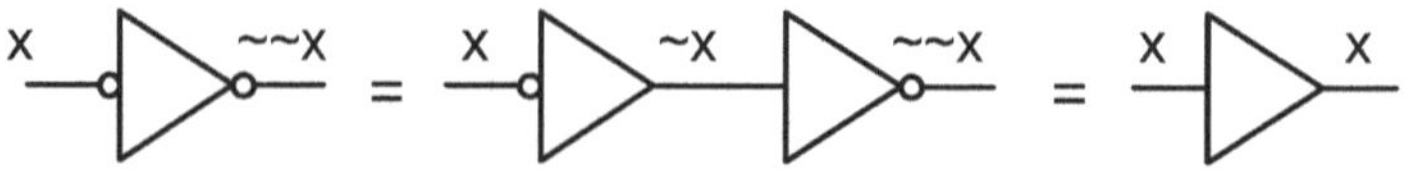

Figure C.10 An alternate representation of an inverter.

From Eqs. (C.1) to (C.3) we see that we can state both forms of De Morgan's theorem as follows:

1. NOT all variables.
2. Change & to | and | to &.
3. NOT the result.

For example,

```
~x | ~y = ~(~~x & ~~y) = ~(x & y)
~(x & y) = ~~(~x | ~y) = ~x | ~y
                                            (C.4)
~x & ~y = ~(~~x | ~~y) = ~(x | y)
~(x | y) = ~~(~x & ~y) = ~x & ~y
```

C.3 Sum of Products Design

We can form a product term, called a *minterm*, for each row of a truth table. The minterm is formed by ANDing together the values associated with each input variable. If the value of the variable in a particular row of the truth table is 1, we include the variable name, such as *x*. If the value of the variable in a particular row of the truth table is 0, we include the negation of the variable name, such as *~x*. Thus, the minterm in row 0 (numbering the rows from 0 to 3) will be *m*0 = *~x & ~y* because both *x* and *y* are 0 in this row as shown in Fig. C.11. Note that the value of minterm *m*0 will be 1 when both *x* and *y* are zero. On the other hand the minterm in row 2 will be *m*2 = *x & ~y*. Note that the value of minterm *m*2 will be 1 when *x* = 1 and *y* = 0. All four minterms are shown in Fig. C.11.

x	y	minterms
0	0	m0 = ~x & ~y
0	1	m1 = ~x & y
1	0	m2 = x & ~y
1	1	m3 = x & y

x	y	z	
0	0	0	
0	1	1	← m1 = ~x & y
1	0	1	← m2 = x & ~y
1	1	0	

z = m1 | m2
= (~x & y) | (x & ~y)

Figure C.11 Sum of products design based on the 1's of the output.

On the right side of Fig. C.11 we show the truth table for the Exclusive-OR function. If we focus on the rows in which the output is 1 (true), we see that the output *z* will be true (1) if either the minterm *m*1 is true *or* if the minterm *m*2 is true. Note that *m*1 will be true if *x* is 0 and *y* is 1 because then *~x & y* will be 1. We can then write the equation for *z* by simply ORing the minterms associated with each output that is 1. From Fig. C.11 we see that

```
z = m1 | m2
  = (~x & y) | (x & ~y)                     (C.5)
```

which is the same as the Exclusive-OR equation we used in Fig. C.6. We will assume the following order of precedence for the logical operators ~, & and |.

1. All ~ operations are done first.
2. All & operations are done next.
3. All | operations are done last.

Equation (C.5) can then be written without the parentheses as

```
z = ~x & y | x & ~y                                    (C.6)
```

ORing all the minterms associated with each 1 in the output column of a truth table leads to a *sum of products* design. The OR operator | is a logical sum and the AND operator & is a logical product. The logical circuit corresponding to Eq. (C.6) is shown in Fig. C.12, which is the same as the one shown in Fig. C.6.

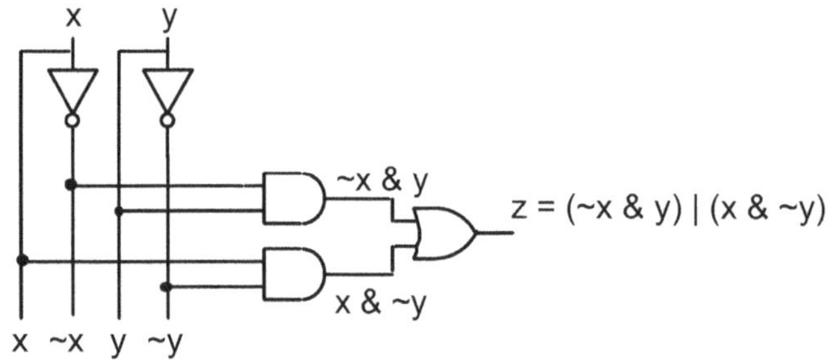

Figure C.12 Logic circuit for the Exclusive-or function.

C.4 Product of Sums Design

Instead of focusing on the 1's in the output column of a truth table, suppose we focus on the 0's, as shown in Fig. C.13. Note that in this case *z* is NOT the minterm *m*0 AND it is NOT the minterm *m*3. What does it mean to be NOT a minterm? From the definitions of minterms in Fig. C.11 and using De Morgan's theorem, we can write

```
NOT minterm m0 = ~m0
               = ~(~x & ~y)
               = ~~(x | y)
               = x | y
```

We call this the maxterm, *M*0. The maxterms for all rows in the truth table are given in Fig. C.14. Note that each maxterm is NOT the corresponding minterm. Use De Morgan's theorem to verify each maxterm expression.

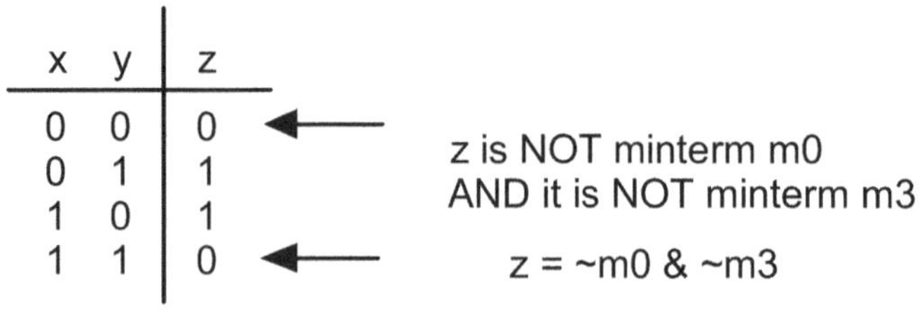

Figure C.13 Focusing on the 0's will lead to a product of sums design.

<table>
<tr><th>x</th><th>y</th><th>minterms</th><th>maxterms</th></tr>
<tr><td>0</td><td>0</td><td>m0 = ~x & ~y</td><td>M0 = ~m0 = x | y</td></tr>
<tr><td>0</td><td>1</td><td>m1 = ~x & y</td><td>M1 = ~m1 = x | ~y</td></tr>
<tr><td>1</td><td>0</td><td>m2 = x & ~y</td><td>M2 = ~m2 = ~x | y</td></tr>
<tr><td>1</td><td>1</td><td>m3 = x & y</td><td>M3 = ~m3 = ~x | ~y</td></tr>
</table>

Figure C.14 A maxterm is NOT the corresponding minterm.

Combining the results in Figs. C.13 and C.14 we see that we can write an equation for *z* as the product (&) of all maxterms for rows in which the output is a zero. Thus, as shown in Fig. C.15, the equation for the Exclusive-OR is

```
z = M0 & M3
  = (x | y) & (~x | ~y)                                   (C.7)
```

which is in the form of a product (&) of sums (|). The logic circuit corresponding to Eq. (C.7) is shown in Fig. C.16. Compare this with the Exclusive-OR circuit shown in Fig. C.12. Note that different logic circuits can perform the identical function.

<table>
<tr><th>x</th><th>y</th><th>z</th><th></th></tr>
<tr><td>0</td><td>0</td><td>0</td><td>← M0 = x | y</td></tr>
<tr><td>0</td><td>1</td><td>1</td><td></td></tr>
<tr><td>1</td><td>0</td><td>1</td><td></td></tr>
<tr><td>1</td><td>1</td><td>0</td><td>← M3 = ~x | ~y</td></tr>
</table>

z = M0 & M3
  = (x | y) & (~x | ~y)

Figure C.15 Product of sums design based on the 0's of the output.

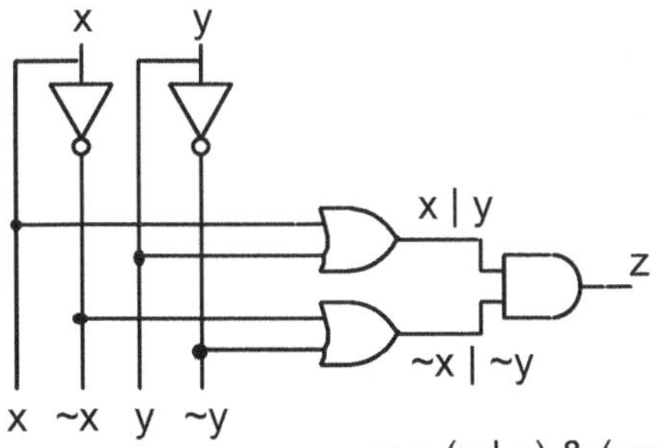

z = (x | y) & (~x | ~y)

<table>
<tr><th>x</th><th>y</th><th>~x</th><th>~y</th><th>x | y</th><th>~x | ~y</th><th>z</th></tr>
<tr><td>0</td><td>0</td><td>1</td><td>1</td><td>0</td><td>1</td><td>0</td></tr>
<tr><td>0</td><td>1</td><td>1</td><td>0</td><td>1</td><td>1</td><td>1</td></tr>
<tr><td>1</td><td>0</td><td>0</td><td>1</td><td>1</td><td>1</td><td>1</td></tr>
<tr><td>1</td><td>1</td><td>0</td><td>0</td><td>1</td><td>0</td><td>0</td></tr>
</table>

Figure C.16 Logic diagram for product of sums Exclusive-OR design.

Appendix D

Boolean Algebra and Logic Equations

George Boole (1815–1864) was a self-taught English logician and mathematician who never went to college. In 1847 he published *The Mathematical Analysis of Logic – Being an Essay Towards a Calculus of Deductive Reasoning* and in 1854 he published *An Investigation of the Laws of Thought.* These works, in which he developed an algebra of logic (now called Boolean algebra), established the relationship between mathematics and logic.

The English philosopher and mathematician Bertrand Russell (1872–1970) and Alfred North Whitehead (1862–1947) published the first of their three-volume *Principia Mathematica* in 1910. This transformed Boolean algebra into a powerful system of *symbolic logic.* In probably the most influential master's thesis ever published, Claude E. Shannon, while a student at MIT in 1938, showed that Boolean algebra, the calculus of symbolic logic, could be applied to relay circuits. He called this switching algebra and showed how these logic equations could allow you to construct circuits to do such things as add and subtract binary numbers.[1] In this book we will use the terms Boolean algebra and switching algebra interchangeably.

In this chapter you will learn

- the basic theorems of Boolean algebra
- how to verify these theorems using a truth table
- how to verify these theorems using Venn diagrams
- how to use Boolean algebra to reduce the complexity of a logic equation
- how to use Karnaugh maps to reduce the number of terms in a logic equation
- the techniques used by computers to minimize the number of terms in a logic equation

D.1 Boolean Theorems

In this section we will study some Boolean algebra theorems and show how you can verify these theorems using Venn diagrams.

---

[1] Claude E. Shannon, "A Symbolic Analysis of Relay and Switching Circuits," Trans. AIEE, Vol. 57, pp. 713-723, 1938. After earning a PhD at MIT Shannon joined Bell Laboratories in 1941 where he became the "father" of modern day information theory.

One-Variable Theorems

Table D.1 shows a list of Boolean algebra theorems involving a single logical variable. All of these expressions are easily verified by showing that each expression holds for values of $x = 0$ and $x = 1$. (See Problems C.10 and C.11.)

<table>
<tr><th colspan="3">Table D.1

One-Variable Boolean Algebra Theorems</th></tr>
<tr><th></th><th>OR-version</th><th>AND-version</th></tr>
<tr><td>Identities</td><td>x | 0 = x
x | 1 = 1</td><td>x & 1 = x
x & 0 = 0</td></tr>
<tr><td>Complements</td><td>x | ~x = 1</td><td>x & ~x = 0</td></tr>
<tr><td>Indempotence</td><td>x | x = x</td><td>x & x = x</td></tr>
<tr><td>Involution</td><td>~~x = x</td><td></td></tr>
</table>

Principle of Duality

In Table D.1 note that the theorems in the AND-version column differ from the theorems in the OR-version column in that | has been replaced with &, all 0's become 1's and all 1's become 0's. This is a general *principle of duality* which holds for any Boolean theorem or identity. It states that any such theorem or identity will remain true if all | and & symbols are interchanged and all 0's and 1's are interchanged. We will see this principle of duality in the two- and three-variable theorems we will look at next.

Two- and Three-Variable Theorems

Table D.2 shows a list of two- and three-variable Boolean algebra theorems. They are divided into five groups and we will look at each group separately. Note in each group that the (b) version is just the dual of the (a) version.

<table>
<tr><th colspan="3">Table D.2

Two- and Three-Variable Boolean Algebra Theorems</th></tr>
<tr><td>Commutative Laws</td><td>(1a)
(1b)</td><td>x | y = y | x
x & y = y & x</td></tr>
<tr><td>Associative Laws</td><td>(2a)
(2b)</td><td>x | (y | z) = (x | y) | z
x & (y & z) = (x & y) & z</td></tr>
<tr><td>Distributive Laws</td><td>(3a)
(3b)</td><td>x | (y & z) = (x | y) & (x | z)
x & (y | z) = (x & y) | (x & z)</td></tr>
<tr><td>Unity</td><td>(4a)
(4b)</td><td>(x & y) | (~x & y) = y
(x | y) & (~x | y) = y</td></tr>
<tr><td>Absorption</td><td>(5a)
(5b)</td><td>x | (x & y) = x
x & (x | y) = x</td></tr>
<tr><td></td><td>(6a)
(6b)</td><td>x | (~x & y) = x | y
x & (~x | y) = x & y</td></tr>
</table>

Commutative Laws

The commutative laws, (1a) and (1b) in Table D.2, should be obvious. They are true because the output values *z* in the two middle rows of the truth tables in Figs. C.2 and C.3 are the same. That is, the same value of *z* occurs regardless of the order of *x* and *y*.

Associative Laws

The associative laws, (2a) and (2b) in Table D.2, state that for three (or more) input variables the order in which an OR or AND operation of all variables is carried out is immaterial. We can always prove any of these theorems by making a truth table and showing that the law holds for all possibilities. This is shown for the associative law, (2a), in Fig. D.1.

<table>
<tr><th>x</th><th>y</th><th>z</th><th>y | z</th><th>x | (y | z)</th><th>x | y</th><th>(x | y) | z</th></tr>
<tr><td>0</td><td>0</td><td>0</td><td>0</td><td>0</td><td>0</td><td>0</td></tr>
<tr><td>0</td><td>0</td><td>1</td><td>1</td><td>1</td><td>0</td><td>1</td></tr>
<tr><td>0</td><td>1</td><td>0</td><td>1</td><td>1</td><td>1</td><td>1</td></tr>
<tr><td>0</td><td>1</td><td>1</td><td>1</td><td>1</td><td>1</td><td>1</td></tr>
<tr><td>1</td><td>0</td><td>0</td><td>0</td><td>1</td><td>1</td><td>1</td></tr>
<tr><td>1</td><td>0</td><td>1</td><td>1</td><td>1</td><td>1</td><td>1</td></tr>
<tr><td>1</td><td>1</td><td>0</td><td>1</td><td>1</td><td>1</td><td>1</td></tr>
<tr><td>1</td><td>1</td><td>1</td><td>1</td><td>1</td><td>1</td><td>1</td></tr>
</table>

Figure D.1 Truth table for the associative law
x | (y | z) = (x | y) | z

A useful graphical technique, called *Venn diagrams*, can be used to verify this and the other theorems in Table D.2.

Venn Diagrams

A *Venn diagram* is a graphical representation of a logical variable or Boolean function. The Venn diagram in Fig. D.2a represents the logical variable *x* as a circle within a unit square. The area inside the circle is thought of as representing a *true* value of *x*, i.e. $x = 1$. The area outside the circle represents values of $x = 0$ or $\sim x$.

If we have two variables, *x* and *y*, then each is represented by a circle. The area inside *x* will represent true values of *x* and the area inside *y* will represent true values of *y*. The area that is outside *x* but inside *y* would represent the Boolean function $\sim x$ & y as shown by the shaded area in Fig. D.2b.

The three logical variables, *x*, *y* and *z* would each have their own circles. The Venn diagrams for the Boolean functions (*x* & *y*), (*x* & *z*), (*y* & *z*) and (*x* & *y* & *z*) are shown in Fig. D.3. Note in each case that the shaded area represents that part of the diagrams in which the *true* values of the variables overlap. This means that *both* (or *all*) variables must be *true* for the AND operation to be true.

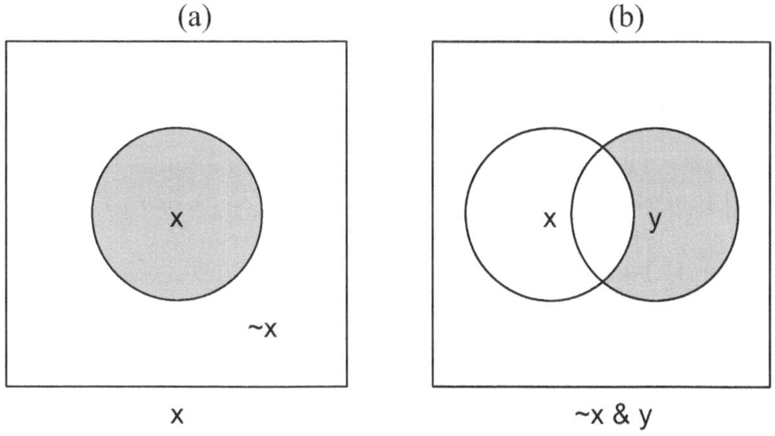

Figure D.2 Venn diagram for logical variable x and Boolean function ~x & y

The validity of the associative law (2b) in Table D.2 should be apparent from the Venn diagrams in Fig. D.3. If you AND the (*y* & *z*) area in Fig. D.3c with *x* you will get the same area (shown in Fig. D.3d) that you get if you AND the (*x* & *y*) area in Fig. D.3a with *z*.

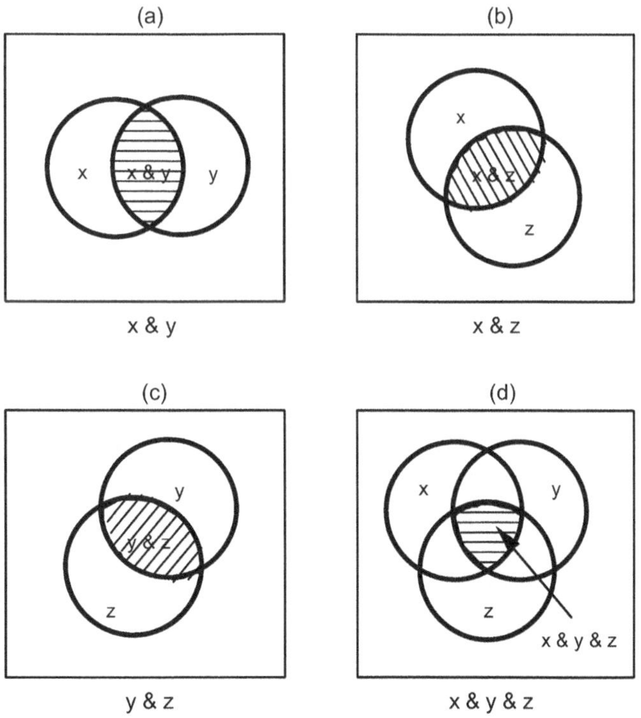

Figure D.3 Venn diagrams for AND operations

The OR operation Venn diagrams are shown in Fig. D.4. Note in this case that the shaded area covers all logical variables because the OR operation is true if either, or both, variable is true. The associative law (2a) in Table D.2 that we proved with the truth table in Fig. D.1 can also be seen from the Venn diagrams in Fig. D.4. It doesn't matter whether we add (OR) x to $(y \mid z)$ in Fig. D.4c or if we add z to $(x \mid y)$ in Fig. D.4a. We will always get the total shaded area shown in Fig. D.4d.

Distributive Laws

The distributive laws (3a) and (3b) in Table D.2 can also be verified using Venn diagrams. Fig. D.11a shows that if we OR x with $(y \& z)$ from Fig. D.10c we get the same shaded area as if we ANDed (found the common overlap of) $(x \mid y)$ from Fig. D.10a with $(x \mid z)$ from Fig. D.10b.

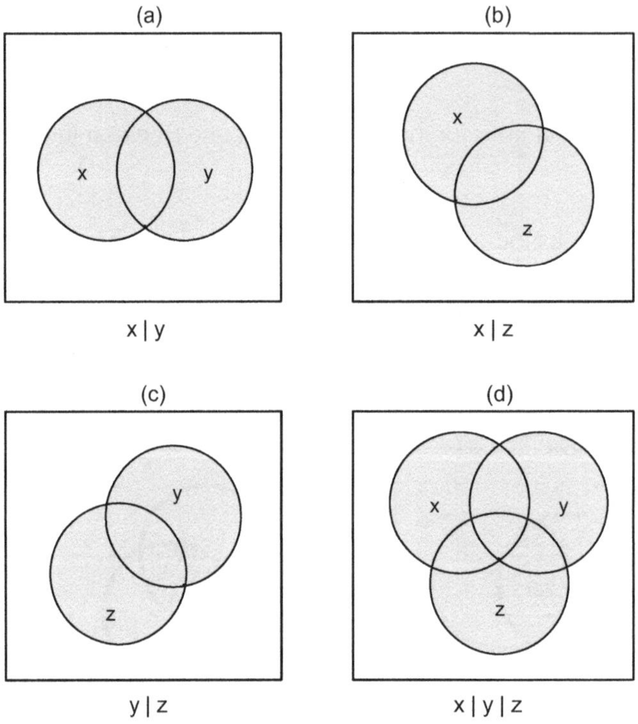

Figure D.4 Venn diagrams for OR operations

Similarly, the shaded area shown in Fig. D.5b can be achieved by ANDing x with $(y \mid z)$ from Fig. D.4c. It can also be obtained by ORing (adding together) the areas $(x \& y)$ from Fig. D.3a and $(x \& z)$ from Fig. D.3b. This verifies the distributive law (3b) in Table D.2.

Note that the distributive law (3b) in Table D.2 is similar to the familiar distributive law in ordinary algebra:

```
x*(y + z) = x*y + x*z
```

In this case the multiplication of x is distributed over the various addition terms. In Boolean algebra, not only is the AND (logical multiplication) operation distributed over the various OR (logical addition) terms as in (3b) of Table D.2, but the OR operation also gets distributed over the various AND terms as in (3a) of Table D.2. The analogous expression in ordinary algebra clearly is NOT true.

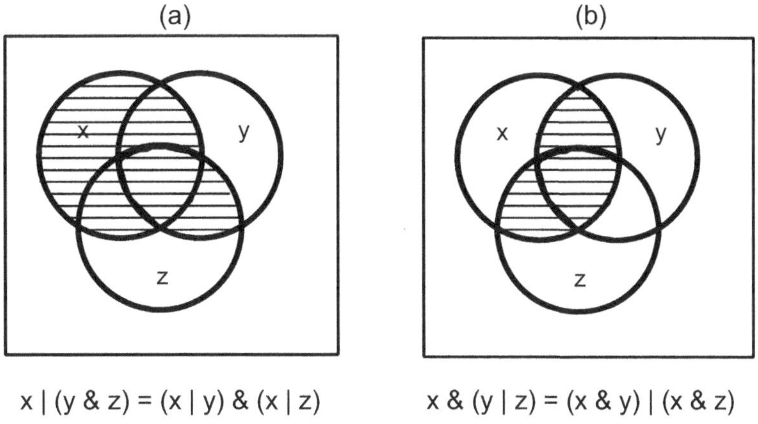

Figure D.5 Venn diagrams illustrating distributive Laws

Unity

The unity theorem (4a) in Table D.2 can be verified in a number of ways. Note that if we add (OR) the shaded area of (*x* & *y*) in Fig. D.9a to the shaded area (~*x* & *y*) in Fig. D.2b we just get *y*.

We can also derive (4a) in Table D.2 from the theorems we already know. We can write

(x & y) | (~x & y)	= (y & x) | (y & ~x)	Table D.2 (1b)
	= y & (x | ~x)	Table D.2 (3b)
	= y & 1	Table D.1
	= y	Table D.1

Similarly, for theorem (4b) in Table D.2 we can write

(x | y) & (~x | y)	= (y | x) & (y | ~x)	Table D.2 (1b)
	= y | (x & ~x)	Table D.2 (3a)
	= y | 0	Table D.1
	= y	Table D.1

Can you see how this last theorem can be verified using Venn diagrams? These unity theorems are sometimes called *combining* theorems.

The form of this unity theorem given in (4a) of Table D.2 is particularly useful in reducing the number of terms in a logic equation. If you find product terms in a sum of products expression that differ only by one variable and its complement, you can "factor out" the common part and the logical sum (|) of the variable and its complement will be 1. We will use this fact in Section D.2 to reduce Boolean functions using a graphical method known as Karnaugh maps.

Absorption

Two forms of the absorption theorem are shown in Table D.2. In (5a,b) the *y* gets absorbed while in (6a,b) the ~*x* gets absorbed. The forms (5a,b) are sometimes called the *covering* theorem where *x* is said to *cover y*.

It is easy to see that if you add the shaded area (*x* & *y*) in Fig. D.3a to *x* in Fig. D.2a you just get *x*. Similarly, if you AND *x* with (*x* | *y*) in Fig. D.4a you still just get *x*.

If you add the shaded areas in Fig. D.2a (*x*) and Fig. D.2b (~*x* & *y*) you get the area (*x* | *y*) in Fig. D.4a, thus verifying the absorption theorem (6a) in Table D.2. Can you see how Venn diagrams can be used to verify theorem (6b) in Table D.2?

D.2 Karnaugh Maps

In Section D.1 we saw how Boolean algebra theorems can be used to reduce the number of product terms in a Boolean function. This will reduce the number of logic elements needed to realize the circuit. In a PLD it is important to reduce the number of product terms so that the Boolean functions will "fit" into the PLD. For example, the GAL 16V8 that we will discuss in Chapter 4 has eight AND gates feeding the OR gate associated with each output pin. This means that we can't program into this PLD a Boolean function that has more than eight product terms. If we have a Boolean function represented by a truth table with many input variables, then every 1 in the output column will represent a product term in the sum of products representation. If there are more than eight of these product terms, we will have to reduce the number to fit it into a GAL 16V8. Computer-based design tools will often reduce the number of product terms for you, but you sometimes need to know how to do it yourself.

There is a simple graphical method that will allow you to reduce the number of terms in a Boolean function that contains two, three or four logical variables. The technique uses what are called Karnaugh maps, or K-maps. For more than four logical variables, the use of K-maps becomes cumbersome and impossible for a large number of variables. Various computer-based tabular methods can be used in these cases.

Two-Variable K-maps

Consider the truth table shown in Fig. D.6a. The sum of products representation of the output function *f* is

```
f = ~x & ~y | x & ~y | x & y                    (D.1)
```

This equation can be written in the following *sum of minterms* notation:

```
f(x,y) = m0 | m2 | m3
```
$$= \sum(0,2,3) \qquad \text{(D.2)}$$

From our Boolean algebra theorems, we can reduce this as follows:

```
f = ~x & ~y | x & ~y | x & y
  = (~x | x) & ~y | x & y          distributive – Table D.2(3b)
  = ~y | x & y                     complement – Table D.1
  = ~y | x                         absorption – Table D.2(6a)
```

Note that this is just the product of sums solution of the truth table in Fig. D.6.

This result is easily obtained by inspection of a two-variable K-map. A K-map is an alternate representation of a truth table in the form of a box of 2 x 2 squares as shown in Fig. D.6. The possible values of *x* (0,1) are used as labels for the rows and the possible values of *y* (0,1) are used as labels for the columns. Thus, each square represents one of the minterms shown in the truth table in Fig. D.6a. The logical values of the output function *f* are inserted in the appropriate minterm square as shown in Fig. D.6b. Note that exactly the same information is contained in Figs. D.6a and D.6b.

minterm	x	y	f
m0	0	0	1
m1	0	1	0
m2	1	0	1
m3	1	1	1

(a)

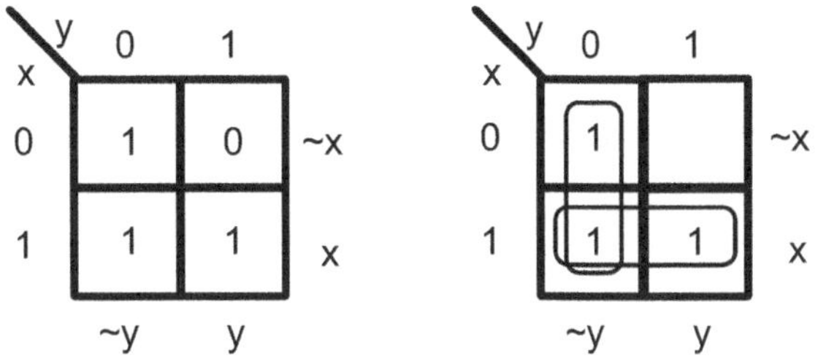

(b) (c)

Figure D.6 A two-variable Karnaugh map

We will refer to the upper left square as the minterm *m0* square, or the 00 square. Similarly, the upper right square is the minterm *m1*, or 01, square; the lower left is the minterm *m2*, or 10, square; and the lower right square is the minterm *m3*, or 11, square.

In a K-map, we normally only show the 1's of the output function and leave the squares containing a zero blank as shown in Fig. D.6c. Now note that if a 1 occurs in two adjacent squares, then two minterms in the sum of products expression will have a variable and its complement in these two terms. We will then be able to eliminate this variable as we've seen above.

For example, consider the two vertical 1's that we have circled in the left hand column of Fig. D.6c. The two minterms associated with these 1's are

```
m0 | m2 = ~x & ~y | x & ~y
        = (~x | x) & ~y
        = 1 & ~y
        = ~y
```

Therefore, these circled 1's contribute the value *~y* to the output *f*. This is easy to see from Fig. D.6c. Anytime the circled 1's cover both the 0 and 1 of a variable, that variable can be eliminated. In this case the circled 1's cover both *~x* and *x* and therefore *x* can be eliminated. This leaves only *~y* because the circled 1's are in the 0 column of *y*. Note that we have labeled this column *~y* at the bottom of the map.

Another set of circled 1's is shown on the bottom row of the map in Fig. D.6c. Note that this circled set covers both *y* and *~y* so that *y* can be eliminated. Since it is in the 1 row of *x*, this just leaves *x*. Therefore, the Boolean function *f* can be written as

```
f = ~y | x
```

as we found above.

The rule for using a 2-variable K-map is to circle all adjacent 1's and then read the result as described above. Note that the circled set of 1's can overlap. If the 1's are not adjacent, e.g. if you have only minterms *m0* and *m3*, then you circle each individual 1 and no reduction is possible.

The original sum of products solution for *f* was

```
f = ~x & ~y | x & ~y | x & y
```

and includes three product terms. By using a K-map we have reduced it to the simple expression *~y* | *x*.

Three-Variable K-maps

K-maps are most useful for problems containing three or four logical variables. The basic ideas described for the two-variable case still apply. For three variables, *x*, *y* and *z*, we draw the K-map as shown in Fig. D.7.

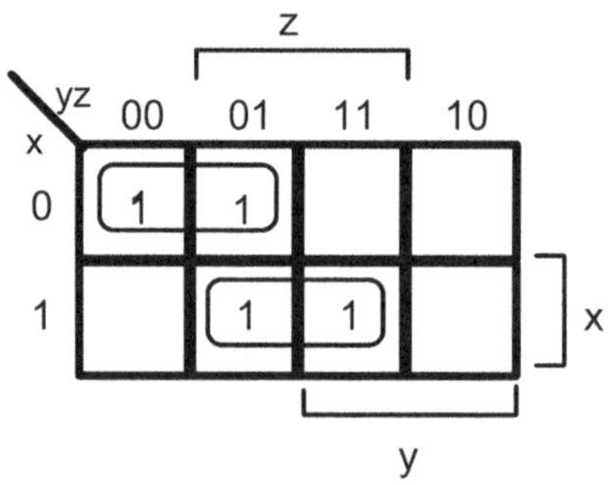

f = ~x & ~y | x & z

Figure D.7 A three-variable K-map

This map has two rows and four columns which gives us eight squares corresponding to the eight minterms of a three-variable function. The rows correspond to the *x* values of 0 and 1 just as in the two-variable case. But now we need to include both variables *y* and *z* in the column labels. Each column is labeled with two binary values. The first is the *y* value and the second is the *z* value. We must include all four possibilities (00 01 10 11). But note that we do *not* write them in this order. Rather we switch the order of the last two and order them (00 01 11 10). The reason we do this is that we must change only one variable at a time (from 1 to 0 or from 0 to 1) in moving from one column to the next. Note that when we do this the two middle columns correspond to $z = 1$ (*true*) while the last two columns correspond to $y = 1$ (*true*).

We circle adjacent 1's just as we did in the two-variable case to eliminate variables. For example, the function shown in Fig. D.7 is given by the four sum of products terms

```
f = ~x & ~y & ~z | ~x & ~y & z | x & ~y & z | x & y & z
```

We can reduce this to two terms by circling the two adjacent sets of 1's as shown in Fig. D.7. The top pair will eliminate the *z* because it covers both a 0 and 1 of the *z*. What's left is a 0 for the *x* and a 0 for the *y* and therefore the remaining term is *~x* & *~y*. Similarly, the bottom pair of 1's covers a 0 and 1 of the *y* and a 1 for the *z* and *x*. Thus, the remaining term is *x* & *z*. The function *f* can then be reduced to

```
f = ~x & ~y | x & z
```

You should always circle as many pairs of 1's at a time as possible. For example, in Fig. D.8 you would circle all four 1's in the bottom row. This will eliminate both *z* and *y* and leave only *x*. The other circled pair eliminates *x*, leaving *y* & ~*z*. This function is therefore written in reduced form as *f* = *x* | *y* & ~*z*.

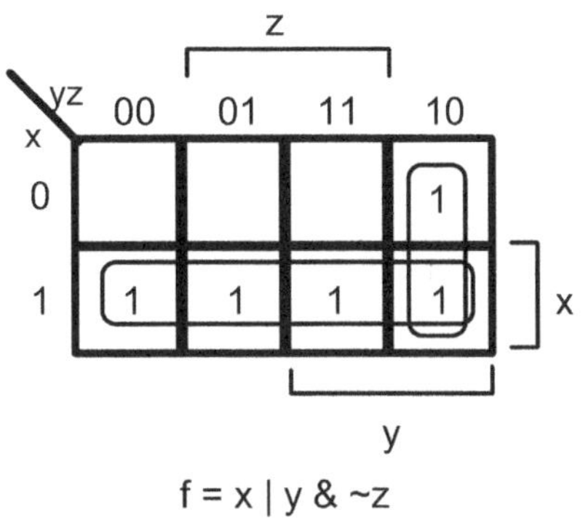

Figure D.8 The circled sets should be a large as possible

If four 1's occur in a large square, rather than a row, you circle the large square as shown in Fig. D.9. This large circled square will eliminate both *x* and *y*, leaving only *z*. The circled pair in the upper left eliminates *z*, leaving ~*x* & ~*y* while the circled pair in the lower right eliminates *z*, leaving *x* & *y*. Thus, the reduced function is

```
f = z | ~x & ~y | x & y
```

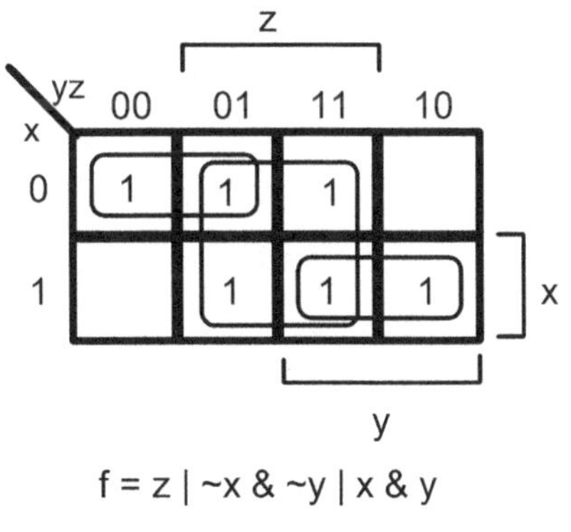

Figure D.9 Four 1's in a square pattern should be circled

You can think of the right-most column as wrapping around to be connected to the left-most column. Therefore, a 1 in the left-most column is "adjacent" to a 1 in the same row of the right-most column. These should be circled as shown in Fig. D.10. This circled pair will eliminate *y*, leaving ~*x* & ~*z*. The lower circled pair in Fig. D.10 contributes *x* & *z* to the function, whose reduced form is then

```
f = ~x & ~z | x & z
```

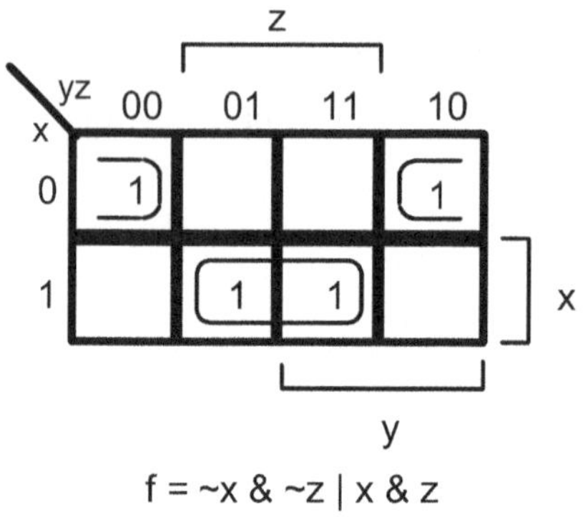

Figure D.10 The left-most column is "adjacent" to the right-most column

As a final example, note in Fig. D.11 that you should circle as many pairs of 1's at a time as you can (without including any zeros), even when they overlap other pairs. In this case the reduced form of the function

```
f = ~z | y
```

has only two terms, each containing a single variable.

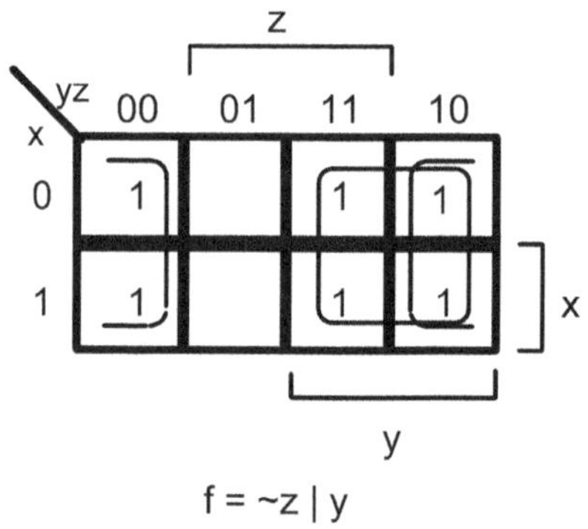

Figure D.11 Make the circled sets of 1's as large as possible

Four-variable K-maps

A Boolean function made up of the four variables *w*, *x*, *y* and *z* can be represented by a four-variable K-map as shown in Fig. D.12. This is a natural extension of the three-variable case where the four rows correspond to the two variables *w* and *x* and the four columns correspond to the variables *y* and *z*. Note that the order of the row labels (00 01 11 10) must be the same as the columns, so that only one variable changes state in going from one row to the next. Again the right-most column is "adjacent" to the left-most column, and the top row is "adjacent" to the bottom row. You can therefore connect 1's at opposite ends of the map within the same circled set.

Fig. D.12 shows how each of the four circled sets contributes to the reduced function

```
f = ~w & y & ~z | ~w & x | x & y | w & ~y & z
```

Note that the original sum of products function would have contained *nine* product terms (one for each 1 in the map).

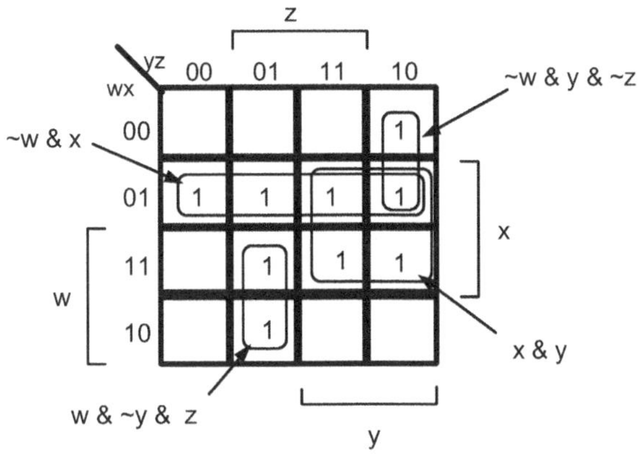

f = ~w & y & ~z | ~w & x | x & y | w & ~y & z

Figure D.12 Example of a four-variable K-map

The minterms associated with each square in a 4 x 4 K-map are shown in Fig. D.13.

wx \ yz	00	01	11	10
00	0	1	3	2
01	4	5	7	6
11	12	13	15	14
10	8	9	11	10

Figure D.13 Minterms in a four-variable K-map

Thus, the function

$$f(w,x,y,z) = \Sigma(0,1,2,3,5,7,8,10,14,15)$$

would have the K-map shown in Fig. D.14. Note that the four corners of a 4 x 4 K-map are adjacent and can therefore be circled. From the K-map in Fig. D.14 the reduced logic equation is

```
f = ~w & z | w & x & y | ~x & ~z
```

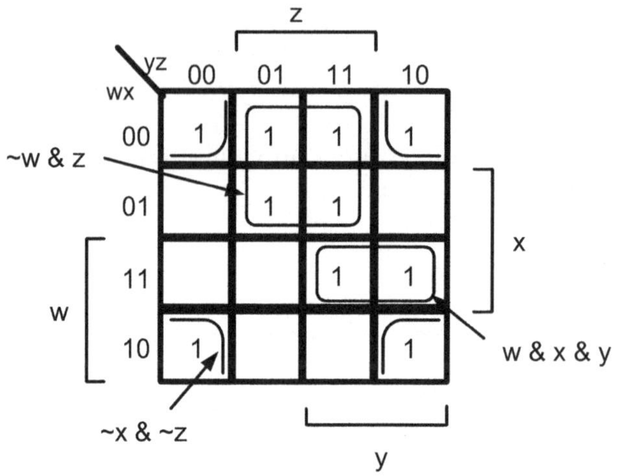

Figure D.14 The four corners of a K-map can be circled

Appendix E

VHDL Quick Reference Guide

<table>
<tr><th>Category</th><th>Definition</th><th>Example</th></tr>
<tr><td>Identifer Names</td><td>Can contain any letter, digit, or underscore _
Must start with alphabetic letter
Can not end with underscore or be a keyword
Case insensitive</td><td>q0
Prime_number
lteflg</td></tr>
<tr><td>Signal Values</td><td>'0' = logic value 0
'1' = logic value 1
'Z' = high impedance
'X' = unknown value</td><td></td></tr>
<tr><td>Numbers and Bit Strings</td><td><base>#xxx#
B = binary
X = hexadecimal
O = octal</td><td>35 (default decimal)
16#C# = "1100"
X"3C" = B"00111100"
O"234" = B"010011100"</td></tr>
<tr><td>Generic statement</td><td>Associates an identifer name with a value that can be overridden with the generic map statement</td><td>generic (N:integer := 8);</td></tr>
<tr><td>generic map</td><td>Assigns a value to a generic parameter</td><td>generic map (N => 16)</td></tr>
<tr><td>Signals and Variables Types</td><td>signal (used to connect one logic element to another)
variable (variables assigned values in process)
integer (useful for loop control variables)</td><td>signal d : std_logic_vector(0 to 3);
signal led: std_logic;
variable q: std_logic_vector(7 downto 0);
variable k: integer;</td></tr>
<tr><td>Program structure</td><td>library IEEE;
use IEEE.STD_LOGIC_1164.all;

entity <identifier> is
    port(
        <port interface list);
end <identifier>;

architecture <identifier> of
        <entity_name> is
begin
  process(clk, clr)
  begin
      {{concurrent_statement}}
  end<identifier>;</td><td>library IEEE;
use IEEE.STD_LOGIC_1164.all;

entity Dff is
    port(
        clk : in STD_LOGIC;
        clr : in STD_LOGIC;
        D : in STD_LOGIC;
        q : out STD_LOGIC);
end Dff;

architecture Dff of Dff is
begin
  process(clk, clr)
  begin
     if(clr = '1') then
        q <= '0';
     elsif(rising_edge(clk))then
        q <= D;
     end if;
  end process;
end Dff;</td></tr>
<tr><td>Logic operators</td><td>not
and
or
nand
nor
xor
xnor</td><td>z <= not y;
c <= a and b;
z <= x or y;
w <= u nand v;
r <= s nor t;
z <= x xor y;
d <= a xnor b;</td></tr>
</table>

VHDL Quick Reference Guide (cont.)

Arithmetic operators	+ (addition) - (subtraction) * (multiplication) / (division) (not synthesizable rem (remainder)	count <= count + 1; q <= q - 1;
Relational operators	=, /=, >, <, >=, <=	**if** a <= b **then** **if** clr = '1' **then**
Shift operators	shl (arg,count) shr (arg,count)	c = shl(a,3); c = shr(a,4);

process	[<id>] **process**(<sensitivity list>)
	{{process declaration}}
	begin
	{{sequential statement}}
	end process [<id>]
		process(a)
		variable j: integer;
		begin
		j := conv_integer(a);
		for i **in** 0 **to** 7 **loop**
		if(i = j) **then**
		y(i) <= '1';
		else
		y(i) <= '0';
		end if;
		end loop;
		end process;

if statement	**if**(expression1) **then**
	{{statement;}}
	{{**elsif** (expression2) **then**
	{{statement;}} }}
	[[**else**
	{{statement;}}]]
	end if;
		if(clr = '1') **then**
		q <= '0';
		elsif(clk'event **and** clk = '1') **then**
		q <= D;
		end if;

case statement	**case** expression **is** ((**when** choices => {sequential statement;}})) {{ … }} **when others** => {sequential statement;}} **end case;**	**case** s **is** **when** "00" => z <= c(0); **when** "01" => z <= c(1); **when** "10" => z <= c(2); **when** "11" => z <= c(3); **when others** => z <= c(0); **end case**;

for loop	**for** identifier **in** range **loop**
	{{sequential statement}
	end loop;
		zv := x(1);
		for i **in** 2 **to** 4 **loop**
		zv := zv **and** x(i);
		end loop;
		z <= zv;

Assignment operator	:= (variable) <= (signal)	z := z + x(i); count <= count + 1;
Port map	instance_name component_name **port map** (port_association_list);	M1 : mux21a **port map**(a => c(0), b => c(1), s => s(0), y => v);

Index

Made in the USA
Las Vegas, NV
10 November 2026

57843106R00111